ONE HEALTH
and the Politics of
COVID-19

ONE HEALTH

and the Politics of COVID-19

Laura H. Kahn, MD, MPH, MPP

JOHNS HOPKINS UNIVERSITY PRESS BALTIMORE

Printed in the United States of America on acid-free paper
9 8 7 6 5 4 3 2 1

Johns Hopkins University Press
2715 North Charles Street
Baltimore, Maryland 21218
www.press.jhu.edu

Library of Congress Cataloging-in-Publication Data

Names: Kahn, Laura H., author.
Title: One Health and the politics of COVID-19 /
Laura H. Kahn, MD, MPH, MPP.
Description: Baltimore : Johns Hopkins University Press,
[2024] | Includes bibliographical references and index.
Identifiers: LCCN 2023050486 | ISBN 9781421449326
(paperback ; alk. paper) | ISBN 9781421449333 (ebook)
Subjects: MESH: COVID-19 | One Health | Biomedical
Research—ethics | Animals | Environment | Politics
Classification: LCC RA644.C67 | NLM WC 506 |
DDC 362.1962/4144—dc23/eng/20240416
LC record available at https://lccn.loc.gov/2023050486

A catalog record for this book is available from the British
Library.

*Special discounts are available for bulk purchases of this book.
For more information, please contact Special Sales at
specialsales@jh.edu.*

Contents

Preface

I dedicate this book to my One Health Initiative colleagues, past and present.

In 2002, when I began my career as a policy researcher at Princeton University, I discovered that physicians and veterinarians rarely if ever communicated, collaborated, or cooperated with each other. This is extremely problematic because combined efforts from these two fields are key to combatting zoonotic diseases, which can easily be transmitted from wild or domesticated animals to humans. As I noted in the article "Confronting Zoonoses, Linking Human and Veterinary Medicine," which appeared in the April 2006 issue of the Centers for Disease Control and Prevention's *Emerging Infectious Diseases*,[1] the medical and veterinary communities were not always so isolated. At the end of the nineteenth century, for example, there was much overlap between the two professions. Drs. Rudolf Virchow and William Osler, who came to be considered the "fathers" of pathology and modern medicine, respectively, both worked closely with veterinarians. Virchow, a German physician, was able to gain significant insight into the zoonotic disease trichinosis by examining thin slices of animal flesh with a microscope; this research led him to establish important food safety policies and procedures. Osler studied under Virchow before returning to Canada to establish the field of veterinary pathology in North America.[2] The subsequent separation of the two professions was unfortunate because zoonotic pathogens such as coronaviruses do not necessarily see a distinction between humans and other animals.

Dr. Bruce Kaplan, a retired veterinarian living in Sarasota, Florida, contacted me after reading my article. He is a former epidemic intelligence officer with the Centers for Disease Control and Prevention who has previously served as a public affairs specialist with the US Department of Agriculture's Office of Public Health and Science, among his many other career hats. Bruce asked me what my plans were after publishing the piece. I didn't know. He suggested that we advocate for physicians and veterinarians to work together, so we set an example by collaborating ourselves—a physician and veterinarian team. Bruce promoted our unusual partnership by sending emails to his many contacts about the need for veterinarians to work with physicians to prevent the spread of zoonotic pathogens from animals to humans.

In February 2007, I met Dr. Thomas P. Monath at an American Society for Microbiology Biothreats meeting, where he was a keynote speaker. Monath is a Harvard-trained physician-virologist, the former chief of the vaccine division at the US Army Medical Research Institute of Infectious Diseases (USAMRIID), and a past president of the American Society of Tropical Medicine and Hygiene. I took Tom's card after his speech and subsequently sent him an email describing the need for physicians and veterinarians to work together. To my delight and surprise, he agreed with our mission, and joined Bruce and me to spread the word. He sent emails detailing our ideas to his many Rolodex contacts around the world. Establishing a One Health Initiative website was Tom's brainchild; on October 1, 2008, the website went live.

Others joined our team, notably:

Jack Woodall, PhD (1935–2016), an entomologist, virologist, and cofounder of the Program for Monitoring Emerging Diseases (ProMED -mail).

Lisa Conti, DVM, MPH (1963–2020), a career Florida state government employee, whose last position before her untimely passing from cancer was as the deputy commissioner and chief science officer of the Florida Department of Agriculture and Consumer Services.

Drs. Thomas Yuill, Helena Chapman, and Craig Carter joined the OHI team in May 2019. Ms. Becky Barrentine joined in May 2020, and Mr. Richard Seifman joined in August 2022.

Thomas M. Yuill, PhD, is professor emeritus at the Nelson Institute of Environmental Studies, School of Veterinary Medicine, University of Wisconsin. He is an expert on wildlife diseases and emerging viral diseases affecting humans and domestic animals.

Helena J. Chapman, MD, MPH, PhD, serves as a professorial lecturer in the Department of Environmental and Occupational Health at the Milken Institute School of Public Health at George Washington University. She also works as an associate program manager for health and air quality applications for the NASA Applied Sciences Program.

Craig N. Carter, DVM, PhD, served in the US Air Force and Army for forty years and has been deployed to or consulted in over thirty-five countries. He received the Bronze Star for commanding the first Army Reserve Veterinary unit in Afghanistan and retired as a full colonel in 2009. He is currently the director of the University of Kentucky Veterinary Diagnostic Laboratory.

Becky Barrentine, MBA, has worked in many corporate business roles, including marketing, communications, administration, accounting, and finance, in a variety of industries, nonprofits, and Fortune 50 firms. She is a founding partner of Crozet BioPharma.

Richard Seifman worked as a senior health advisor with the World Bank and a senior foreign service officer with the US Department of State. He is a board member of the United Nations Association of the National Capital Area and a senior columnist with *Impakter*, an online publication dedicated to bringing innovative policy ideas to the world.

While the term "One Health" is relatively new, the concept is ancient, and it is understood by Indigenous peoples around the world. The global One Health movement began organically, with many of the earliest groups forming in the early to mid-2000s. Our One Health Initiative team always worked pro bono to promote the concept.

This book was written for a broad audience. Medical, veterinary medical, and public health professionals, as well as interested policy makers, students, and general readers, should find it informative and useful. I tried to keep the technical terms to a minimum, but in some cases, their use is essential.

Since 2012, when I published "Going Viral" in the *Bulletin of the Atomic Scientists*, I have been warning about the dangers of gain-of-function research.[3] From a public health perspective, the risks of creating deadly pathogens in laboratories clearly outweighed the benefits.

Some pieces of information, such as the story about the development of mRNA vaccines, are important but tangential to the main flow of the text. These pieces are included in the appendixes and endnotes.

This book is a snapshot of the COVID-19 pandemic. New information is being generated all the time. My goal is to inform readers about the history, science, and politics of coronaviruses and coronavirus research. Learning from history is essential to prevent us from repeating the same mistakes. Coronaviruses should never be underestimated.

ONE HEALTH
and the Politics of
COVID-19

Introduction

The COVID-19 pandemic began as a mystery. The larger story of the connections between coronaviruses, humans, animals, environments, and ecosystems must be examined to put the pandemic into historical context and to help us understand our predicament. One Health is the concept that human, animal, environmental, and ecosystem health are linked. This concept provides a useful framework for examining these connections and will be used, chapter by chapter, to do a deep dive into the history, science, and politics of coronaviruses and coronavirus research.

The microbes that live in us and on us constitute our microbiomes and are as important for our health and well-being as our lungs, kidneys, or intestines. Animals and plants have their own microbiomes, too. Microbes that cause diseases are called pathogens, and those that spread between animals and humans are called zoonoses. Over half of human pathogens have zoonotic origins, and the process of them *emerging* and spreading from animals into human populations is not new.[1] For example, the measles virus emerged from the rinderpest virus, a deadly pathogen of cattle, around the sixth century BCE.[2,3] What's new is our ability to identify, track, and respond to them.

Some animals are natural hosts of zoonotic pathogens. The known host animals for coronaviruses are bats, birds, and rodents.[4] Coronaviruses mutate frequently, enabling them to infect many species.[5] Since the early twentieth century, they have caused numerous deadly animal disease outbreaks called epizootics. If epizootics involve food animals, a common strategy to stop disease spread involves mass culling, but rigorous biosafety, biosecurity, vaccinations, and antimicrobials could provide additional control measures. For example, an avian influenza epizootic of millions of chickens that began in early 2022 resulted in the culling of more than 44 million laying hens in the United States, leading to a national egg shortage and a sharp increase in egg prices.[6]

There is a vaccine for avian influenza, but it hasn't been used largely because of cost and trade considerations. In Europe, policy makers and farmers have been rethinking poultry vaccinations as the avian influenza virus devastates European flocks. While vaccinations wouldn't stop infections, instead potentially allowing the virus to spread silently from animals that are not obviously sick, they would reduce severe disease and the need to cull flocks. In China, vaccinations against avian influenza were mandated in 2017, resulting in zero spillover events in humans. This was a big win for public health.[7]

Until 2003, coronaviruses caused only mild upper respiratory tract infections in humans and were not a medical concern. The situation changed in 2002 when the severe acute respiratory syndrome (SARS) virus (a.k.a. SARS-CoV-1) emerged in the Guangdong Province of southern China. This natural spillover event was associated with the sale of exotic animals in live animal markets and their subsequent butchering and consumption. The virus caused a deadly epidemic spanning more than thirty countries, infecting thousands, and killing more than eight hundred people.[8]

In 2012, another deadly coronavirus emerged. This time, it appeared in Jordan and Saudi Arabia. Called the Middle East respiratory syndrome (MERS), the virus was associated with dromedary camels, which transmitted the virus to humans. Appearing to be deadlier than SARS-CoV-1, the MERS coronavirus (MERS-CoV) spread primarily in the Arabian Peninsula, but it also spread via a traveler to cause a sizeable outbreak in the Republic of Korea.[9]

In late 2019, a coronavirus named SARS-CoV-2 set off the COVID-19 pandemic. The pandemic caused global societal and economic upheavals, infecting tens of millions and killing millions of people. SARS-CoV-2 emerged in Wuhan, the most populous city in the Hubei Province of central China. In addition to housing a large live animal market, the city is also the home of the Wuhan Institute of Virology (WIV), which contains the nation's only biosafety level (BSL) 4 laboratory as well as three BSL-3 and twenty BSL-2 laboratories.[10] Concerns about Chinese biosafety practices in high containment laboratories existed before the pandemic.[11] For example, in 2004, poor biosafety practices in a Beijing virology laboratory led to an escape of SARS-CoV-1, causing a local outbreak.[12] (Of course, no laboratory is immune to human error.[13])

On February 19, 2020, twenty-seven prominent scientists published a letter in the *Lancet* declaring solidarity with their Chinese scientific and medical counterparts and condemning the conspiracy theories suggesting that the pandemic originated from a laboratory. Based on the analyses of the virus's genome by scientists from multiple countries, they concluded that the coronavirus came from wildlife like many other emerging pathogens. Backing their position were the presidents of the US National Academies of Sciences, Engineering, and Medicine.[14]

Many scientists argued for the natural origins of the virus. In March 2020, five prominent virologists published a letter in *Nature Medicine* giving their opinions on the virus, acknowledging the unique features of its spike protein. They stated that given the high rates of genetic variation in coronavirus spike proteins these unique features would likely be discovered in other coronaviruses even though no animal coronaviruses identified so far were sufficiently similar to serve as the direct progenitors of SARS-CoV-2. They concluded that it was "improbable that SARS-CoV-2 emerged through laboratory manipulation."[15]

Nevertheless, calls for a transparent and independent investigation into the origins of the virus began. In November 2020, Dr. David A. Relman, a professor of medicine, microbiology, and immunology at Stanford University School of Medicine, wrote in the *Proceedings of the National Academy of Sciences* that it was critical to understand the chain of events leading to the pandemic to prevent it from happening again. He stressed that conflicts of interest by researchers, administrators, and policy makers must be revealed and addressed to achieve political neutrality and scientific balance. Discussions of risky science, Relman argued, were necessary. Otherwise, the deliberative process would not be credible, trustworthy, or effective.[16]

In the United States, conservative political leaders politicized the virus origin controversy. During the initial years of the pandemic, President Donald J. Trump called the coronavirus the "Chinese Virus" and "Kung Flu," setting off a political backlash against his remarks.[17] Paul Gosar, representative from Arizona, wrote on Twitter that he and his senior staff were under self-quarantine after being exposed to the "Wuhan Virus."[18]

Unfortunately, the rhetoric appeared to set off violence against Asian Americans and hindered open and honest debate about the virus's origins. President Trump was accused of inciting racist attacks. He dismissed the concerns that his language contributed to the surge of verbal and physical assaults around the country.[19]

In response to the antagonistic rhetoric, the mainstream media downplayed the "lab leak" theory. For example, the *New York Times* never published a four-thousand-word article by its science writer Donald G. McNeil Jr. that examined the complicated nature of the scientific evidence being pushed by allies of President Trump.[20] The "lab leak" theory became another toxic conspiracy theory.

On February 27, 2022, the *New York Times* published an article titled "New Research Points to Wuhan Market as Pandemic Origin." It cited two studies, both published on Zenodo, a general-purpose open-access repository operated by the European Organization for Nuclear Research (CERN), which had been designed for researchers to post their work that had not yet been peer reviewed.[21] Michael Worobey, an evolutionary biologist at the University of Arizona, coauthored both studies, which analyzed maps of the animal market stalls, social media activity, and viral genes and concluded that the Huanan market was the epicenter of the pandemic.[22,23] Worobey had previously published a paper in *Science* in November 2021, asserting the same conclusion, and had been cited by the *New York Times*.[24]

On social media, particularly on Twitter, the discourse veered toward support for the lab leak hypothesis. A group of individuals, calling themselves DRASTIC, an acronym for Decentralized Radical Autonomous Search Team Investigating COVID-19, searched for clues about the origins of the virus. To them, the scientific establishment and mainstream media were behaving unscientifically, squelching debate that the virus was anything but natural.[25] For example, the *New York Times* did not cite a study by George Gao, a virologist, director of the Chinese Center for Disease Control and Prevention, and member of the Chinese Academy of Sciences, that was also undergoing peer review around the same time as Worobey's papers, in which he found zero out of 457 samples taken from 18 species of animals in the Huanan market collected in early 2020 that tested positive for the SARS-CoV-2 virus.[26]

The history of coronavirus research involves scientific inquiry, technological advances, and integrity, but it also includes scientific hubris. Ultimately, it is a story about humans and our relationship to microbes.

As this book is being written, the COVID-19 pandemic remains ongoing. We must learn to live sustainably in our microbial world. The first step is recognizing that our lives are connected with the lives of other species. In other words, we must look beyond ourselves and approach our health and well-being in a systematic way, using an interdisciplinary approach called One Health. This is an approach that has been understood by many of the world's Indigenous peoples whose interests reflect the realities of their daily lives.[27]

While the One Health concept recognizes that human, animal, plant, environmental, and ecosystem health are linked, in general, it's not how academia, industry, or governments approach health and well-being. Instead, their primary concern rests solely with humans. For example, most government disease-surveillance systems collect data on humans, not animals.[28] Government disease-surveillance systems that do collect data on animals focus mostly on food-producing animals. However, as we will see in this book, disease surveillance of *all* animals is essential for ensuring global health. Companion animals might serve as reservoirs of antimicrobial resistant bacteria, and wildlife might serve as hosts for zoonotic pathogens.

To examine coronaviruses, this book uses a One Health framework, represented as a multidimensional matrix tool.[29] Dimension One represents the essential One Health factors: humans, animals, plants, environments, and ecosystems. Humans can be stratified by gender, age, or other relevant characteristics, such as disease status. Animals and plants can be either domestic, wild, or both.

Environments encompass the abiotic (e.g., soil, water, air) aspects of defined geographic areas, whereas ecosystems involve the biotic (e.g., microbial, flora, fauna) interactions within defined geographic areas. Environments and ecosystems can be indoors or outdoors in urban, suburban, rural, or natural settings. Dimension Two is divided according to levels of complexity and provides scale: microbial or cellular, individual, and population levels. Dimension Three involves political, social, and economic factors. In this book, the political, social, and economic factors of coronavirus research will be briefly discussed. (Please see appendix 1a to visualize the complete matrix tool.)

Chapter 1 examines the history and science of coronaviruses in animals, and chapter 2 explores coronaviruses in environments and ecosystems. Chapter 3 discusses the history of coronaviruses in humans. Chapters 4 and 5 delve into the molecular biology and research conducted on coronaviruses and the issues of biosafety, biosecurity, and bioethics. Chapter 6 concludes with policy recommendations.

Coronaviruses are part of life on our planet. The COVID-19 pandemic has caused an unprecedented global public health crisis. We must stop the risky activities that promote their emergence and spread, otherwise we should assume that more coronavirus pandemics are in our future.

Dimensions One and Two

	One Health and Complexity Factors	Microbial/ Cellular	Individual	Populations
CHAPTER ONE	**Domestic and Wild Animals**	X	X	X
CHAPTER TWO	**Environments and Ecosystems**			X
CHAPTER THREE	**Humans**	X	X	X
CHAPTER FOUR	**Molecular Biology of Coronaviruses**	X		

Time: Chronological

Domestic and Wild Animals

Chicken flocks, cattle herds, swine herds, and other domesticated livestock exemplify the livelihoods of farmers and ranchers. These food producers conduct careful surveillance of their animals' health and seek professional veterinary medical assistance when diseases appear in order to minimize economic losses and to protect animal welfare. As a result, the earliest recognition of a newly emerging pathogen occurred in flocks of sick chicks in the 1930s in the United States.

A Novel Virus Emerges

In April 1930, Drs. Arthur Frederick Schalk and Merle C. Hawn, veterinarians working with the North Dakota Agricultural College in Fargo, North Dakota, found themselves inundated with dead and dying baby chicks delivered both by mail and in person by hatchery men and poultry keepers. The chicks that were still alive were gasping for air.

In their twenty years of field and laboratory experience, the veterinarians had never seen anything like it. Nor could they find descriptions of the disease in the veterinary medical literature. They tentatively concluded that they were witnessing a new respiratory disease of chicks.

The disease spread quickly. Outbreaks were occurring in Illinois, Iowa, Louisiana, Mississippi, Nebraska, Ohio, and South Dakota. In their own state of North Dakota, tens of thousands of chicks were dying and causing enormous economic losses for the poultry industry.

The signs of the disease were dramatic, appearing approximately three to four days after the chicks hatched. The baby birds would exhibit intermittent signs of listlessness and distress. Occasionally, they would liven up and peck at their feed. Between days five and nine, gasping signs would begin. Eventually, the birds' wings would sag, their feathers ruffle, their eyes close, and they would hold their heads close to their bodies.

They would assume a crouched position and intermittently stretch their heads upward and forward while gasping for air. Some would emit a distinctive low sonorous rattling sound. They would become emaciated and exhausted and subsequently die.

While conducting necropsies (animal autopsies) on hundreds of chicks, Schalk and Hawn expected to find lesions in the birds' respiratory tracts because they thought they were dealing with laryngotracheitis, a highly contagious herpesvirus infection causing inflammation of the larynxes and tracheas of poultry. Instead, they found the bronchi and lungs partially or completely filled with serous-mucoid exudate, a clear, thick fluid. In other words, the birds drowned in their own fluids. The mortality rate ranged from 40 to 90 percent. Schalk and Hawn dubbed this illness "infectious bronchitis of baby chicks."[1,2]

Veterinarians L. D. Bushnell and C. A. Brandly also witnessed the disease during the 1930 hatching season while working at the Kansas Agricultural Experiment Station in Manhattan, Kansas. Like their colleagues in North Dakota, they found fluid-filled lungs and tracheas while conducting necropsies on hundreds of chicks. To determine the etiological agent, they conducted experiments and found that the disease could be transmitted from sick to healthy chicks using bacteriologically sterile filtrates of kidney, liver, and spleen tissues and blood. They concluded that the cause of the gasping disease must be a filterable virus.[3]

The Discovery of Viruses

At that time during the early 1930s, little was known about viruses. Their discovery had been a relatively recent event using crude technologies such as porcelain filters to isolate them. The first virus to be discovered was the tobacco mosaic virus, a plant pathogen.

In 1879, Adolf Mayer, the director of the Agricultural Experiment Station and professor of botany at the Agricultural School in Wageningen, Netherlands, had been studying a disease of tobacco plants that he called "tobacco mosaic disease" because of the pigmented spots on the diseased plants' leaves. He had been trying to satisfy Koch's postulates, the experimental gold standard for proving the microbial cause of disease, by infecting healthy plants with sap from the leaves of diseased plants. But he was unable to prove that bacteria caused the disease.[4] He used special filters to process the sap and found that the bacteria-free liquid remained infectious. He shared his findings with Martinus Beijerinck, a chemical engineer and younger colleague at the facility.

Eight years later and more than 1,300 miles away from Wageningen, Dmitri Ivanovsky, a young Russian botanist, also conducted experiments with tobacco mosaic disease using unglazed porcelain filters with tiny pores to filter out bacteria. He concluded that very small infectious agents, smaller than bacteria, were capable of passing through the filters to cause disease. He speculated that bacterial toxins in the sap might be the culprit.

Beijerinck worked for ten years as an industrial microbiologist before returning to academia at the Technical University of Delft and resuming the research on tobacco mosaic disease that Mayer and Ivanovsky had begun. Using two filtering systems—unglazed porcelain and agar, a gelatinous substance made from red algae—he proved that the cause of the plant disease was an infectious agent, incapable of independent growth, requiring the presence of host cells in order to replicate, and much smaller than bacteria. He believed that the new agent was liquid in nature and called it *contagium vivum fluidum*.[5] In Latin, the term "virus" means "liquid poison." Decades later, Wendell M. Stanley, a chemist working in the Rockefeller Institute's laboratories in Princeton, New Jersey, used X-rays to show that the tobacco mosaic virus was a particle, not a liquid, and was composed of protein and ribonucleic acid (RNA).[6,7] The porcelain filters used to discover the tobacco mosaic virus were similar to those used to research the virus killing chicks on American farms.

Avian Infectious Bronchitis Virus

In 1932, Drs. Jerry R. Beach and O. W. Schalm, veterinarians working at the Agricultural Experiment Station at the University of California, Berkeley, confirmed Bushnell and Brandly's findings in Kansas that the respiratory disease of chicks was distinct from laryngotracheitis and was caused by a new virus. (Of note, Beach had also discovered the laryngotracheitis virus of poultry, which was subsequently determined to be a herpesvirus.)[8,9,10]

Five years later, the infectious bronchitis virus was cultivated for the first time in chick embryos by Dr. Fred Beaudette, a veterinarian, and Charles Hudson, a poultry research specialist, at the New Jersey Agricultural Experiment Station.[11] They took infected fluids from sick birds and injected them into fertilized chicken eggs. Some of the chick embryos died but not all of them. They collected fluids from the surviving embryos and injected them into chickens that were known to be immune to the laryngotracheitis virus. The chickens that recovered from the infection produced antibodies capable of neutralizing the new virus.[12] In other words, they had created an antibody treatment for the infectious bronchitis virus.

While the infectious bronchitis virus was the first coronavirus to be discovered in domesticated animals, it would not be the last. Indeed, each succeeding decade since the 1930s experienced outbreaks caused by different coronaviruses infecting domesticated pigs, laboratory mice, turkeys, cats, dogs, cattle, and horses. The coronaviruses infected different organ systems: pulmonary, gastrointestinal, hepatic (liver), lymphatic, splenic, neurologic, and others. In some outbreaks, the viruses were extremely contagious, infecting millions of animals. (A brief timeline of coronavirus outbreaks is available in appendix 1b.)

The Electron Microscope and Coronaviruses

The electron microscope revolutionized virology by allowing researchers to visualize the infinitesimal pathogens causing disease. In 1931, Ernst Ruska built the first electron microscope for his PhD thesis with his advisor Max Knoll, a German electrical engineer and the leader of an electron research group at the Technical College of Berlin. Eight years after developing the electron microscope, Ruska and his colleagues G. A. Kausche and E. Pfankuch used his invention to visualize the tobacco mosaic virus.[13] For his work, Ruska was awarded the Nobel Prize in Physics in 1986.[14]

Donald Berry at Glaxo Research in Middlesex, England, together with a group of pathologists and veterinarian colleagues at Cambridge University, was the first to examine the avian infectious bronchitis virus using an electron microscope. It had a strange spherical appearance with "spikes" on its surface that appeared attached by very narrow necks and sometimes forming bulbous masses at the distal ends.[15] Other viruses shared this appearance, as well as the unusual characteristic of storing RNA as genetic material instead of the more stable deoxyribonucleic acid (DNA) in their capsules. In response to these findings, eight virologists wrote a letter to *Nature* suggesting the name "coronavirus" for this new group of viruses because they looked like solar coronas. Corona means "crown" in Latin.[16] A brief history of human coronavirus discoveries will be discussed in chapter 3.

Coronaviruses as Zoonotic Pathogens

SARS-CoV-1

The 2002–2003 severe acute respiratory syndrome (SARS) spillover event in the Guangdong Province of southeast China shocked the world. Until then, coronaviruses had killed only domesticated animals, not humans. The newly emerged SARS-CoV infected over eight thousand people and killed over eight hundred in thirty-two countries. The pandemic lasted approximately eight months until stringent public health measures finally contained it.[17,18]

The earliest SARS victims were restaurant chefs and food handlers who killed or butchered wild animals for exotic food. Some lived within walking distance to a live animal market. Farmers working with domestic animals did not appear to be at increased risk of exposure to the virus.[19]

Because the earliest SARS patients had connections with killing, butchering, or cooking wild animals, researchers began their search of wild animals to find the source of the virus. One study swabbed oral and anal orifices of wild animals being sold at a live animal market in Guangdong, China, and found SARS-CoV-like viruses from Himalayan masked palm civets, a racoon dog (a type of wild dog native to East Asia), and a Chinese ferret badger (a carnivorous mammal native to Southeast Asia). Two of the viruses obtained from the civets were closely related to the SARS virus. The authors concluded that the live animal markets likely provided a venue for the virus to jump from species to species but cautioned that the animals testing positive were probably not the natural reservoir hosts.[20]

Another study zeroed in on civets. It found that 80 percent of the civets from a live animal market in Guangzhou tested positive for SARS-CoV antibodies whereas virtually none of the civets tested from four different farms in Guangdong Province had them. The civets were likely infected with the SARS-CoV virus in the live animal market, where animals of many different wild and domestic species were caged close together.[21] Adding further credence to this theory was another study that captured and tested wild civets in Hong Kong and found them negative for SARS-CoV-like viruses.[22]

Masked palm civets are catlike mammals that eat fruits, birds, insects, and small vertebrates—in other words, they are omnivores, like humans. They are solitary, nocturnal, tree dwellers found in Africa as well as in South and East Asia. In parts of China, they are considered a culinary delicacy and are cooked in a popular soup known as *dragon, tiger, and phoenix*. Cobras or rat snakes represent the "dragon," civets the "tiger," and chickens the "phoenix." The soup is believed to have health benefits for people suffering from arthritis, poor blood flow, and low libido. Scrapings from civets' perineal glands have a musky fragrance that are used in expensive perfumes and in certain foods to enhance flavor.[23]

During the time that SARS emerged, Dr. Lin-Fa Wang, a molecular biologist and project leader at the Australian Biosecurity Cooperative Research Centre for Emerging Infectious Diseases in Brisbane, Australia,[24] speculated that bats might be the reservoir host species for the novel coronavirus. Bats were known reservoir hosts for many deadly zoonotic viruses, including rabies and the Nipah and Hendra viruses that had recently emerged in East Asia and Australia. As reservoir host animals, bats rarely displayed signs of illness from these deadly viruses. In addition, bat meat and bat guano were eaten as food and used in traditional Chinese medicines, respectively, in southern China. Bat guano contains many pathogenic bacteria, fungi, and viruses.[25] Dr. Wang and a team of researchers traveled to the Guangdong, Guangxi, Hubei, and Tianjin Provinces in China and trapped 408 bats from nine different species and six genera. They took oral and fecal swabs as well as blood specimens and tested them for SARS-CoV antibodies using different methods in separate laboratories in Geelong, Australia, and Wuhan, China.[26]

They hit pay dirt with horseshoe bats.[27,28,29]

Three communal, cave-dwelling species of the genus *Rhinolophus*, also known as horseshoe bats, had high titers of SARS-CoV antibodies in the blood specimens. These bats were found in the Guangxi and Hubei Provinces, which are over 600 miles (1,000 kilometers) apart. Because they were widely distributed geographically, demonstrated high seroprevalence rates of antibodies to the SARS-like coronavirus, and harbored coronaviruses with greater genetic diversity than those isolated from masked palm civets or humans, the authors concluded that horseshoe bats were likely the natural reservoir hosts of the virus.[30]

From the virus's perspective, bats (order Chiroptera, "hand wing") make excellent reservoir hosts. First, there are many of them. Of the 4,600 known mammalian species, over 20 percent are bats. They are an ancient species, appearing around 52 to 50 million years ago, coincident with a rise in insect and plant diversity as well as increased global temperatures.[31] Second, bats fly, which allows them to travel long distances, helping viruses to spread. They are geographically located on all continents except Antarctica, fill many ecological niches, and consume a wide variety of foods, including fish, fruit, insects, pollen, and blood. Third, they have long life spans—up to thirty-five years—and are very social, often roosting in enormous colonies, which helps viruses evolve between animals. Fourth, they hibernate during winter months, helping to maintain viruses during severe weather conditions.[32] Most importantly, the immune systems of bats allow many viruses to persist in their bodies without evidence of disease.

Bats are difficult to study, which hinders scientific investigations to understand how their immune systems work. One theory regarding bats' toleration of viral infections is that flight increases their metabolisms and elevates their body temperatures analogous to fevers in other mammals.[33] And indeed, there is evidence suggesting that the evolution of flight might have influenced their limited inflammatory responses to viruses, which also coincidentally might have contributed to their longevity.[34,35]

The abundance and diversity of viruses in bats, termed the bat "virome," is a rapidly growing area of scientific interest with important public health implications. In 2006, 66 viruses from 18 viral families had been detected in bats.[36] By 2014, an online database listed over 4,100 bat-associated viruses from 23 viral families found in 196 bat species across 69 countries worldwide. Single-stranded RNA viruses were the most common viruses found in bats, and of these, coronaviruses constituted almost 80 percent of them.[37] The database is updated every two months with increasing numbers of viruses.[38]

MERS-CoV

Almost a decade after the SARS epidemic, another coronavirus spread to humans in the Middle East. This coronavirus appeared to be deadlier than SARS, with a reported mortality rate of 35 percent versus 9.6 percent, respectively.[39] Fortunately, it was not as contagious as SARS. In June 2012, the Middle East respiratory syndrome coronavirus (MERS-CoV) was identified for the first time in a sixty-year-old man in Saudi Arabia.[40,41] Searching for the source of the new pathogen, scientists collected specimens from domesticated livestock and found that only dromedary camels (one-humped camels) tested positive for MERS-CoV antibodies. Throughout the Middle East, North and East Africa, and parts of Asia, camels tested positive.[42] There was evidence that the virus had been circulating for a long time because archived blood specimens from camels dating back at least three decades also tested positive for MERS-CoV antibodies.[43,44]

Dromedary camels were domesticated sometime between 2000 and 1000 BCE, a relatively recent phenomenon compared to other animals such as sheep and cattle, which were domesticated around 10,500 to 9,000 years ago, respectively. Because of their unique adaptations to arid, hot environments, camels served as important means for trade and travel before modern transportation became widely used throughout the Levant, the Arabian Peninsula, and the Sahara. Camels are strictly herbivores and graze on grasses, grains, and oats. In the twenty-first century, camels remain integral components of Arabian culture for tourism and racing, as well as for meat, milk, and fiber.[45] As of 2019, there were approximately 35 million camels living primarily in North and East Africa, the Middle East, and Central and South Asia; 95 percent were dromedary camels as opposed to Bactrian (two-humped) camels.[46]

There are similarities and differences between masked palm civets and camels as intermediate hosts for SARS and MERS, respectively. Masked palm civets aré either wild caught or farm raised and are sold in live animal markets in China. They have been farm raised for fur since the 1950s and have become increasingly popular as exotic food since the 1980s. In 2003, approximately forty thousand civets were raised on six hundred farms in China.[47] While SARS-CoV-1 infected civets, signs of illness in these animals had not been reported. In contrast to civets, camels have been domesticated for thousands of years. MERS-CoV generally causes mild upper respiratory tract infections in camels, but they can also exhibit fevers, coughing, sneezing, loss of appetite, or no signs of illness at all.[48]

As with SARS-CoV, bats were suspected as the reservoir hosts for MERS-CoV. To investigate, Dr. Ian Lipkin, an infectious disease physician-scientist at Columbia University in New York City, and his colleagues collected blood specimens, throat swabs, rectal swabs, and fecal pellets from ninety-six bats representing seven different species from a variety of locations, including an abandoned date palm orchard, near where the first MERS victim lived in Bisha, Saudi Arabia. Over a quarter of the fecal specimens contained a wide variety of coronaviruses. One specimen from an Egyptian tomb bat (*Taphozous perforates*) contained a piece of coronavirus RNA that was identical to a segment of MERS-CoV found in the MERS victim. Unfortunately, the entire virus could not be recovered. Lipkin attributed the inability to sequence the entire coronavirus RNA to a customs official who had opened the transport container preserving the specimens on dry ice in order to inspect it. Over two days, the specimens deteriorated at room temperature.[49,50]

MERS-CoV has been determined to represent a novel coronavirus lineage, or clade, named 2c, which is distinct from the SARS-CoVs (a.k.a. Sarbecoviruses). A South African study examining fecal pellets from sixty-two bats representing thirteen different species found five coronaviruses. One specimen identified as PML/2011 from an adult female *Neoromicia* cf. *zuluensis* bat differed from MERS-CoV by one amino acid (0.3 percent) in a gene fragment. This finding made it a closer relative to MERS than a previous virus isolated from *Pipistrellus* bats that differed by 1.8 percent.[51,52]

While bats are the acknowledged reservoir host species that spread MERS-CoV to camels, it's unknown exactly how and when they first infected camels. The virus took several decades to spread from camels to humans. At some point, the virus's spikelike proteins evolved the ability to unlock the receptors on human cells, especially those lining the respiratory tract.[53] A likely mode of spread from camels to humans was through respiratory droplets since sequences of the virus were detected in nasal swab specimens from camels. Occupational exposure to camels, including training and herding camels, milking camels, not washing hands after handling camels, and cleaning farm equipment, was found to be a risk factor for seropositivity to MERS-CoV. Drinking camel milk, particularly if it wasn't pasteurized, was another potential risk factor.[54,55,56]

SARS-CoV-2

Seventeen years after SARS and seven years after MERS, a third coronavirus, SARS-CoV-2, emerged in fall 2019 in Wuhan, Hubei Province, China. Yet again, the world watched in shocked disbelief as the Coronavirus Disease 2019 (COVID-19) rapidly spread. Unlike the previous two coronavirus epidemics, the direct animal source of COVID-19 could not be determined.

Chinese officials initially reported that the virus came from the Huanan (South China) Seafood Market, where seafood and farmed and wild animals were sold.[57] The largest live animal market of its kind in central China, covering an area the size of nine football fields and housing over one thousand stalls, was closed on January 1, 2020, to be cleaned and disinfected in response to the crisis. Some of the initial patients were market employees, stall owners, or visitors to the market.[58] Health officials from the Chinese Center for Disease Control and Prevention reportedly went to the market to collect specimens and found evidence of the virus in 33 out of 585 samples taken from people and from animal stalls.[59]

Several months later, however, Gao Fu, the director of China's Center for Disease Control and Prevention, announced on China Global Television Network that the coronavirus found in samples of the market's local environment had not been detected in any of the animals at the market.[60] Gao and his team had collected environmental samples inside the Huanan market in early 2020; 64 out of 828 (7.7 percent) samples had tested positive for the human virus HCoV/Wuhan/IVDC-HB-01 and had shared a nucleotide identity of 99.9 percent with it. But *no* virus was detected in 457 animal samples taken from 18 species of animals, including 10 stray cats, 27 cat feces, 1 dog, 1 weasel, and 10 rats in the market.[61]

Adding further evidence that the market was not the initial source of the virus, a *Lancet* paper describing the earliest hospitalized patients with pneumonia in Wuhan reported that fourteen out of the forty-one

(34 percent) patients had no direct exposure to the market. Indeed, the patient with the earliest symptoms, beginning on December 1, 2019, had no known exposure to the market and no links to any of the other cases.[62]

Some scientists speculated that the original virus spilled over into humans in China as early as October 2019.[63,64] In other words, it was more likely that someone brought the virus *into* the animal market rather than becoming infected *from* it. The reservoir host animal was assumed to be a bat; however, rodents in China have also been found to harbor many different coronaviruses.[65]

In the case of COVID-19, the assumption of bats as the reservoir hosts appeared to be correct. Chinese researchers led by Zheng-Li Shi, a virologist at the Wuhan Institute of Virology, reported that the viral genomes (the complete set of the viruses' genetic instructions) from five patients during the early stages of the pandemic were 96 percent identical to the genome of a bat coronavirus (Bat-CoV-RaTG13) obtained six years earlier from horseshoe bats (*Rhinolophus affinis*) in an abandoned mineshaft in Mojiang County in the Yunnan Province of China. But the bat and human coronaviruses did not match perfectly, and the mineshaft was located 1,000 miles southwest of Wuhan. In other words, Bat-CoV-RaTG13 would have had to travel 1,000 miles and mutate 4 percent of its genome to become SARS-CoV-2.[66,67,68]

While coronaviruses are capable of spreading directly from bats to humans,[69] the arguments against Bat-CoV-RaTG13 as the direct predecessor of SARS-CoV-2 prompted scientists to continue searching for an intermediate animal host. Their search led to Malayan pangolins (*Manis javanica*), which are nocturnal, scale-covered, termite- and ant-eating mammals. In China, pangolin meat is extremely popular and considered a culinary delicacy. Pangolins' keratin scales (made from the same protein found in hair, nails, claws, horns, and hooves) are components of traditional Chinese medicines. They have been used to treat crying children and women "possessed by the devil."[70]

In high-end Chinese restaurants, pangolins are hammered unconscious and slaughtered in front of the diners to assure them that the meat is fresh. The blood is drained and given to the customers to take home. Pangolin meat is either boiled, braised, steamed, or stewed. Baby pangolins are served whole in pangolin fetus soup. The enormous demand for pangolins has driven them to the verge of extinction, and they are listed as one of the most endangered species by the Convention on Illegal Trade in Endangered Species of Wild Fauna and Flora (CITES).[71,72]

Three studies examined the possible role of pangolins as intermediate hosts. The first study examined smuggled pangolins seized from August 2017 to January 2019 by Guangxi customs officers in southern China. Scientists collected samples of lung, intestine, and blood specimens from eighteen pangolins to test for evidence of coronaviruses. In six of the forty-three samples taken, they found coronaviruses closely related to SARS-CoV-2. But there were key differences between the pangolin-derived coronaviruses and SARS-CoV-2, particularly in the parts of the spike protein that binds to human cells. Despite the differences, the study authors urged the banning of pangolin sales in live animal markets and extreme caution whenever handling them because they might have played a role in coronavirus spread to humans.[73]

The second study looked at smuggled pangolins seized by customs officers in 2019 in China's Guangdong Province. Three pangolins were examined. All three were sick with severe respiratory illnesses and died despite rescue efforts by a wildlife rescue center. It's unclear if the animals were sick from coronaviruses or from another pathogen. The coronaviruses isolated were more closely related to bat coronaviruses than to SARS-CoV-2. The study authors concluded that SARS-CoV-2 was unlikely to have evolved directly from pangolin coronaviruses.[74]

The third study examined lung tissues from twenty-five smuggled Malayan pangolins in a wildlife rescue center. Some of the animals were sick with pneumonia. One of the pangolins harbored a coronavirus that was very similar genetically to Bat-CoV RaTG13, and it shared spike protein sequences with SARS-CoV-2. Despite the similarities, there were enough differences between the viruses that the authors concluded that the pangolin-derived coronavirus was unlikely to be linked to the COVID-19 pandemic. Their conclusion: humans should leave highly endangered pangolins alone regardless of the findings because they might serve as coronavirus transmitters in future pandemics.[75]

None of the studies conclusively determined that pangolins were the intermediate hosts. Aside from pangolins, snakes and turtles were briefly considered as possible intermediate hosts but were quickly dismissed. Chickens, ducks, geese, mice, rats, monkeys, sheep, cows, horses, pigs, dogs, cats, ferrets, camels, rats, and many wild animals tested negative for SARS-CoV-2 antibodies, negating their potential roles as intermediate hosts.[76] Tissue samples from wild animals such as raccoon dogs, hog badgers, Siberian weasels, and hundreds of bats in and around Wuhan were sampled. None of these samples yielded SARS-CoV-2 or its progenitor virus. However, newly discovered alphacoronaviruses were identified from the raccoon dogs that shared a 97.9 percent genome sequence similarity with coronavirus strains found in domestic dogs in the United Kingdom.[77]

Zoonotic pathogens such as SARS-CoV-2 infect many animals, but it is not always a one-way street. Sometimes, humans infect the animals. Indeed, as the COVID-19 pandemic spread, humans began infecting a wide variety of animals, including pet cats and dogs; zoo tigers, lions, leopards, and gorillas; laboratory ferrets, hamsters, monkeys, and mice; and farmed mink.[78,79] Wild white-tailed deer populations became infected with the virus in several US states.[80]

Farms provided opportunities for cross-species transmission. Mink farms in countries including the Netherlands, Spain, Sweden, Italy, the United States, and Denmark became infected with SARS-CoV-2.[81] In April 2020, two mink (*Neovision vision*) farms in the Netherlands were the first to report SARS-CoV-2 in their animals. A subsequent investigation of sixteen Dutch mink farms with SARS-CoV-2-positive animals found that in some cases employees infected the mink, while in others, the mink infected the employees. How the mink on farms without SARS-CoV-2-positive employees became initially infected is unknown. But mink can infect each other as well.[82]

In Denmark, SARS-CoV-2 mutations developed in mink and people, raising concerns that the mutated viruses might reduce vaccine effectiveness. Danish prime minister Mette Frederiksen decided to cull 17 million mink to reduce the risks.[83] The slaughter of the country's entire mink population resulted in enormous trenches filled with decomposing bodies, some of which resurfaced. The media called them "zombie minks." The bodies posed contamination risks to drinking water supplies. And since the prime minister made the cull decision without Parliament's approval, the entire operation was deemed illegal.[84]

A SARS-CoV-2 variant named Omicron might have come from mice. Chinese researchers at the University of Chinese Academy of Sciences in

Beijing suggested that the highly contagious Omicron variant, first reported in South Africa on November 24, 2021, originated in mice after being infected by humans, sometime in mid-2020, before jumping back into humans in late 2021. According to the researchers, the Omicron variant had 45 unique point mutations compared to the collective 6,986 point mutations of the other variants that had spread in humans worldwide. The researchers did not specifically state if they thought the cross-species transmission occurred in a laboratory.[85]

The coronavirus saga began in the early twentieth century with outbreaks in baby chicks. New epizootics affecting many animal species were detected in virtually every succeeding decade. In the twenty-first century, the scientific community has been studying these wild animal reservoirs to better understand their ecologies. Chapter 2 will describe some of these efforts. Since human activities such as the trade and consumption of wild animals continue, future coronavirus spillover events from animals to humans should be anticipated.

Environments and Ecosystems

Background

Unfathomable numbers of viruses inhabit Earth. They exist in environments and ecosystems wherever life is found, and while not alive, viruses possess enormous biological and genetic diversity. They are technically not alive because they do not eat, eliminate wastes, or function as complete microorganisms in that they only reproduce by infecting target host cells.

The term "virosphere" is used to describe the parts of the planet affected by viruses. In the world's oceans alone, there are an estimated 4×10^{30} viruses—a number larger than the number of visible stars in the universe.[1] More than 90 percent of the biomass in the world's oceans are microorganisms, primarily prokaryotes (e.g., bacteria, archaea, and blue-green algae), and viruses kill approximately 20 percent of them daily.[2] Understanding the effects of environmental conditions on pathogenic viruses and the microbial ecosystems they inhabit is essential for developing effective disease prevention strategies.

For example, ocean temperatures affect host and virus interactions. A four-year study done in coastal Maine sampled oral-cloacal swabs and fecal specimens from wild ducks and gulls and found that influenza A virus stability and infection capabilities in the samples were influenced by ocean temperatures. The researchers speculated that other environmental factors, including water salinity, invertebrate and plankton population concentrations, ultraviolet light, and manmade pollutants, might also affect viral disease transmission capabilities.[3]

Viruses can float in the atmospheric boundary layer (the lowest part of Earth's atmosphere) for thousands of kilometers and rain down by the billions per square meter each day.[4] To do this, they either attach to dust particles of terrestrial origin or to organic marine aerosols.[5]

There is evidence that they remain infectious after long-distance aerosol transport.[6] In other words, viruses appear capable of seeding across vast distances to find new environments and new ecosystems to inhabit.

Bottom line: viruses should not be underestimated.

In this book, environments are defined as the abiotic (i.e., soil, water, air) aspects of geographic areas. Ecosystems are defined as the biotic interactions between microbes, flora, and fauna within geographic areas. Environments and ecosystems can include indoor and outdoor settings. Outdoor settings can range from urban, suburban, and rural areas to undeveloped, natural regions.

Coronavirus Ecology

In response to avian influenza epidemics in the late 1990s and the SARS epidemic in the early 2000s, the United States Agency for International Development established the Emerging Pandemic Threats (EPT) program in 2005 to assist developing countries in detecting, controlling, and preventing zoonotic diseases with pandemic potential.[7] Within EPT, the Pandemic Influenza and Other Emerging Threats unit established the PREDICT project along with other projects.[8] In its first five years, 2009–2014, PREDICT involved a research consortium across twenty countries in Africa, Asia, and Latin America that sampled thousands of bats, rodents, nonhuman primates, and humans to better understand coronavirus diversity and ecology. Over 19,000 animals and humans were tested for coronaviruses, of which bats represented 64 percent of the total sample. Of the study subjects that tested positive, 8.6 percent were bats (1,065/12,333) compared to 0.2 percent that were non-bat species (17/6,859). The researchers estimated an average number of 2.67 coronaviruses per bat species, or to extrapolate, somewhere between 1,200 to 6,000 coronaviruses exist in the world's total bat populations. Regions with more diverse bat species, particularly in Africa and Asia, had more diverse coronaviruses, the majority of which have yet to be discovered. Importantly, the study found that bat guano, urine, saliva, and nasal secretions were important environmental sources for coronaviruses.[9]

Coronavirus Transmission

Coronaviruses use a variety of methods to spread to new hosts, so determining the primary means of transmission is critical for containing disease spread. Zoonotic spillover of coronaviruses from animals to humans was discussed in the previous chapter. This section will focus on human-to-human viral transmission.

People produce respiratory droplets and bioaerosols through coughing, sneezing, singing, talking, and sometimes by simply breathing.[10] However, confirming airborne transmission of pathogens can be difficult because of the technical challenges involved in sampling bioaerosols.[11] Coronaviruses can be found in urine, stool, and wastewater.[12] Certain individuals are more infectious than others and cause "superspreading" events that infect many people.[13] In some settings, such as day care centers, schools, and nursing homes, viral transmission by fomites transmitted by contaminated objects such as table surfaces and doorknobs theoretically could be possible.[14] Environmental conditions such as temperature, humidity, airflow, and ventilation might also influence viral spread.[15]

Quantifying how fast an epidemic is spreading provides an important public health metric and communication tool. The basic reproductive number, R, is the number of successful secondary cases caused by one infected primary case. R_0 (R naught) is calculated at the beginning of an epidemic when a population is completely susceptible to the pathogen. It is a summary measure and is typically estimated using mathematical models. The models use data based on three primary parameters: the duration of contagiousness of an infected individual, the likelihood that the infected individual will infect a susceptible contact, and the rate of contacts per the infected individual. Other variables, such as how a microbe spreads (e.g., droplet versus airborne), population density, indoor versus outdoor settings, and seasons (e.g., winter versus summer), should also be considered.[16]

R_t is another quantity that can be used during an epidemic, usually after a population has developed some resistance either through disease recovery or vaccination. Its value is typically lower than R_0 because there are fewer susceptible people during the later stages of an epidemic. Nevertheless, R_t can be as challenging to estimate as R_0.

The higher the R_0, the more contagious the pathogen. For example, measles is one of the most contagious viruses in the world with an R_0 commonly cited as ranging from 12 to 18, but it can have even wider values.[17] This means that one sick person with measles can infect, on average, twelve to eighteen people.

When R_0 is greater than 1, the epidemic will spread. When R_0 is less than 1, the epidemic will die out. Therefore, the goal of public health is to get R_0 to be less than 1, meaning that one sick person won't infect anyone else. Getting R_0 less than 1 can be achieved through various methods, including social distancing, personal protective equipment, personal hygiene practices, environmental disinfection, air dilution, quarantining healthy contacts, isolating sick individuals, and if available, vaccination.[18]

SARS-CoV-1 Viral Transmission

Once the SARS coronavirus spilled over from animals into humans, it spread person to person primarily through infectious respiratory droplets into the mucous membranes of people's eyes, noses, and mouths. Maximum virus excretion from the respiratory tract peaked around day ten from illness onset, but the period of maximum infectiousness varied by individual. Highly symptomatic individuals were responsible for most of the disease spread, largely through close contacts in household or hospital settings, but as previously mentioned, some individuals caused superspreading events. Heavy virus-laden droplets traveled only a few meters before dropping to the ground as opposed to airborne transmission, which can infect an entire room through a cough or sneeze. Fecal-oral transmission rates were unknown but theoretically possible. A World Health Organization (WHO) consensus meeting estimated the SARS R_0 to be approximately 3.0.[19]

Hong Kong was particularly hard-hit by the SARS virus with an R_0 ranging from 1.7 to 3.6.[20] Environmental factors appeared to play a major role in viral spread. For example, in one large housing complex outbreak, there was evidence of airborne transmission involving more than three hundred people. More than half of the cases occurred in a single building. One "superspreading" individual with very high virus concentrations in their urine and diarrhea visited a middle-floor apartment in the building on two separate occasions, using the toilet each time. When flushing the toilet, the index case produced an infectious bioaerosol plume. The plume entered an air shaft, ascended to the upper floors, infecting some of the residents in the adjacent units, exited the airshaft, and blew on a northeasterly wind to nearby buildings, infecting some of the residents living in apartments facing the wind. The index case also infected others through person-to-person contact and in communal indoor spaces such as staircases and elevators.[21]

Despite the occasional superspreading events, by July 2003, the SARS epidemic was fully contained by simple public health measures.[22] No vaccine had been developed.[23]

One year later, a SARS outbreak resulting from laboratory-acquired infections in two researchers working in Beijing, China, led to eight confirmed cases and the need to quarantine hundreds of individuals.[24] The WHO confirmed that breaches of laboratory safety procedures were the probable cause of the outbreak, which resulted in the death of one individual.[25]

MERS-CoV Viral Transmission

For MERS, the R_0 ranged from 0.45 to 3.90 in Saudi Arabia to 8.10 in South Korea.[26,27] In most cases, MERS appeared to spread in health care and household settings from close contact with sick individuals and their respiratory droplets. Health care workers such as nurses were especially vulnerable because of prolonged and repeated exposures to MERS patients. However, because of the high R_0 in South Korea, there was concern that it might spread by airborne and fomite routes.[28]

The South Korean outbreak began when a 68-year-old man with chronic medical conditions returned from a business trip to Saudi Arabia, the United Arab Emirates, and Bahrain. He had no contact with camels and had never visited a hospital. On May 15, 2015, this "superspreading" individual was admitted to Pyeongtaek St. Mary's Hospital with a fever and coughing. He was hospitalized in a room on the eighth floor for two days before being prematurely discharged. While on the eighth floor, he infected twenty-six individuals, including patients on the seventh and eighth floors as well as nurses. The secondary cases infected an additional ten people, making a total of thirty-six cases in this MERS hospital outbreak. Six people died.[29] An investigation of the outbreak inside Pyeongtaek St. Mary's Hospital concluded that the virus spread through close contact with respiratory droplets as well as long-range airborne routes.[30] The index case ultimately survived his infection after receiving intensive care in another hospital, but he inadvertently infected other people after seeking care at other health care facilities.[31]

Since the overall R_0 for MERS was less than 1 in most settings, the virus never achieved sustained human-to-human transmission. However, aerosol transmission did occur in some cases. Risk factors for disease transmission included direct contact with MERS patients, not wearing an N95 respirator, lack of personal protective equipment, and lack of isolating MERS patients in hospital settings. Viable virus was found in air samples taken from rooms of MERS patients as well as on fomites such as bed rails, bedsheets, intravenous fluid hangers, and radiograph devices, which could facilitate disease spread. Another study recovered viable MERS-CoV on plastic and steel surfaces in low temperatures (20°C) and low humidity (40 percent) after forty-eight hours.[32] In general, containing the virus was achieved through standard public health measures.[33] However, sporadic MERS outbreaks have continued, with new mutations resulting in potentially deadlier variant strains.[34]

SARS-CoV-2 Viral Transmission

When COVID-19 first appeared in Wuhan, China, it was easy to assume that its transmission would be like the SARS and MERS epidemics involving primarily droplet and direct contact spread.

That assumption was wrong.

The SARS-CoV-2 virus spread much faster and more efficiently than either of the SARS or MERS viruses. Asymptomatic infected individuals could spread the SARS-CoV-2 virus.[35,36,37] Superspreading events accounted for substantial numbers of cases, and the virus readily spread by long-range airborne transmission, especially in indoor settings.[38]

Basic Reproduction Number

During the early days of the pandemic, estimating R_0 proved to be a challenge since most researchers assumed that only symptomatic individuals could spread the virus, and nobody knew how fast one infected person could infect others.[39] Without accurate, widespread screening tests, it would be virtually impossible to find the answers.

Nevertheless, efforts to estimate R_0 for COVID-19 proceeded.

Evidence of human-to-human transmission appeared very early in Wuhan, China. Chinese epidemiologists analyzed data from 425 laboratory-confirmed cases of "pneumonia of unknown etiology" between December 2019 and January 2020 and found an exponential growth of cases. They estimated that the time from viral exposure to the earliest appearance of symptoms (i.e., the incubation period) was 5.2 days. The outbreak doubling time was 7.4 days. Based on these and other measures, they estimated that R_0 was 2.2 and concluded that since R_0 for the new pathogen was less than 3.0, which was the estimated R_0 for the original 2003 SARS epidemic, the ongoing outbreak should be successfully contained through careful infection control, isolation of patients, and 14-day quarantines of contacts. However, none of the 425 laboratory-confirmed pneumonia cases had included children under the age of 15 years.[40]

A separate investigation of a family cluster infected with the novel coronavirus found that a 10-year-old child with radiological evidence of pneumonia did not have any symptoms. Six family members had radiological evidence of pneumonia, ranging in age from 10 to 66 years. Five adults had symptoms, including cough, diarrhea, sore throat, and weakness. They had traveled to Wuhan from Shenzhen from the end of December 2019 to the beginning of January 2020. Two of them visited a sick family member in a Wuhan hospital. These two individuals might have become infectious before symptom onset because one was shedding the virus without any symptoms. The incubation period appeared to range from three to six days. The virus attack rate within the family was high, around 83 percent.[41] The findings of potential asymptomatic spread had ominous implications.

Postdoctoral researchers and mathematical modelers at the Los Alamos National Laboratory in Los Alamos, New Mexico, also estimated the R_0 of the rapidly spreading coronavirus. They collected 137 confirmed, publicly available case reports from China, focusing on individuals who had traveled domestically to Wuhan, gotten infected, returned home, and became the earliest COVID-19 cases in their home provinces. As a result, much more accurate epidemiologic data was obtained from these individuals compared to the earliest cases in Wuhan. Based on these case reports, the researchers estimated an incubation period of 4.2 days, a doubling time of 2.3 to 3.3 days, and an R_0 of 5.7, much higher than the R_0 from the previous study.[42] Their work suggested a much more rapidly spreading virus than previous reports had indicated.

As the virus spread, many variants appeared, with differing levels of communicability. The Delta variant, also known as B.1.617.2, emerged in India in May 2021 and rapidly spread to other countries.[43,44] Over the succeeding months, it became the dominant global variant, infecting people regardless of their vaccination status. (However, those who were vaccinated were much less likely to develop severe disease or die.) Outbreaks associated with large public gatherings in Massachusetts in July 2021, despite high vaccination rates, prompted the Centers for Disease Control and Prevention (CDC) to issue an alert that all individuals, regardless of vaccination status, should wear face masks in indoor public settings.[45] Tom Wenseleers, a Belgian biostatistician, estimated the Delta variant R_0 to be 6.0 or 7.0, about two to three times more contagious than the original SARS-CoV-2 virus.[46] And the even more contagious Omicron variant has been estimated to have an average R_0 of 9.5 with a range of 5.5 to 24.0.[47]

Indoor Transmission

Indoor environments facilitated greater COVID-19 spread compared to outdoor settings. On January 25, 2020, an 80-year-old passenger on the *Diamond Princess* cruise ship disembarked in Hong Kong to go to a hospital after experiencing several days of COVID-19 symptoms while on board the ship. On February 1, the passenger tested positive for the virus.[48] A total of 3,711 individuals, including 2,666 passengers and 1,045 crew members, were placed in a fourteen-day quarantine onboard the cruise ship docked offshore Yokohama, Japan.[49] After February 5, the passengers were confined to their cabins to reduce viral spread, but one study determined that the peak incidence of disease transmission had already occurred.[50] Many had been exposed at mass gathering events involving dancing, singing, shopping, and watching performances in common recreational areas.[51] Seven hundred and twelve passengers tested positive for COVID-19, of whom 46.5 percent were asymptomatic. Nine people died.[52] The ship became the largest COVID-19 outbreak outside of China and made headlines worldwide. At least twenty-five other cruise ships confirmed positive COVID-19 cases. The returning passengers carried the virus with them to their home countries.[53]

Superspreading Events

Superspreading events involving single individuals infecting up to ten other people propelled viral transmission.[54] Near Antwerp, Belgium, a Christmas festival at an elderly care home led to a COVID-19 outbreak of forty staff members and over one hundred residents, of whom at least twenty-six died. The costumed Santa, who had subsequently tested positive, had unintentionally infected everyone.[55]

In February and March 2020, multiple infected travelers from Europe arrived in the northeastern United States. One epidemiological study analyzed the viral genomes of SARS-CoV-2 in the greater Boston, Massachusetts, area to understand viral transmission, particularly at superspreading events. One such event occurred in late February 2020 at a Biogen conference, a two-day international business meeting, at a Boston hotel. At least one hundred cases were directly associated with the conference. Viral genomes collected from twenty-eight patients who had attended the conference were sequenced, revealing a viral lineage that had been circulating in Europe. The conference attendees had spent hours together inside poorly ventilated rooms without wearing face masks. Viral spread flourished.[56] Two variants, C2416T and G26233T, were associated with the conference and were used to track subsequent viral spread. By early November 2020, the Biogen conference–linked variants had traveled nationally to twenty-nine US states and internationally to countries including Australia, Slovakia, and Sweden. The researchers estimated that by November 2020, these variants would infect hundreds of thousands of people.[57]

Wastewater Surveillance

As the coronavirus overwhelmed beleaguered public health surveillance systems, researchers investigated whether the surveillance of sewage might be useful to track viral spread since shedding of SARS-CoV-2 in feces had been reported.[58,59] Historically, wastewater surveillance of enteric pathogens such as hepatitis A, norovirus, and polio had proven useful to identify viral circulation in populations.[60,61] A team in the Netherlands collected samples of incoming wastewater at different wastewater treatment plants at five cities and the main airport before and after February 27, 2020, the day the first Dutch COVID-19 case was reported. A positive wastewater sample was detected on March 5 in Amersfoort, the Netherlands' fifteenth-largest city, six days before health officials reported two COVID-19 cases there. This finding suggested that the virus had been circulating in the city before Dutch health surveillance systems had detected it. By March 25, all wastewater samples in all sites had tested positive for the virus. The study demonstrated the usefulness of wastewater surveillance for early SARS-CoV-2 detection and its prevalence in populations.[62]

In response, interest in SARS-CoV-2 wastewater surveillance exploded, especially since one treatment plant can capture wastewater from over 1 million people, including individuals with minimal or no symptoms.[63] More than fifty nations began surveillance of wastewater for the appearance of SARS-CoV-2.[64] During the fall 2020 academic period, many colleges in the United States used wastewater surveillance in conjunction with laboratory testing to prevent disease spread on their campuses.[65]

Health officials in the United Arab Emirates tested more than 2,900 untreated municipal wastewater samples from forty-nine sites around Dubai and samples from 198 international commercial flights arriving to the Dubai airport. Overall, the virus positivity rate was 28.6 and 13.6 percent for the municipal and aircraft wastewater samples, respectively. Aircraft from some countries stood out: wastewater from ten out of sixteen flights (63 percent) arriving from Pakistan tested positive. The researchers concluded that wastewater surveillance was more cost effective for COVID-19 prevalence monitoring than nasal swab testing.[66] However, despite the cheaper surveillance costs, wealthy countries were better able to conduct wastewater surveillance because they had municipal sewage treatment facilities. Poor, developing countries with inadequate or nonexistent sewage systems would have greater difficulty implementing such a strategy.[67]

Fecal-Oral Transmission

The presence of SARS-CoV-2 in feces raised the question of the possibility of fecal-oral transmission during the COVID-19 pandemic. Diarrhea had been a well-recognized symptom of COVID-19, and fecal-oral transmission had been seen in animal coronavirus outbreaks. But the role of fecal-oral transmission during the SARS and MERS outbreaks remained more speculative than confirmed. Many studies investigated the possible role of SARS-CoV-2 fecal-oral transmission.

In one study, collaborators in the Netherlands and New Zealand reviewed ninety-five papers published in the medical literature between December 2019 and June 2020 involving different gastrointestinal studies of adults or children with COVID-19. Most of the studies had been done in China. Ninety-six percent of the studies found SARS-CoV-2 genetic material (RNA) in stool specimens, especially in patients with gastrointestinal symptoms. Six patients were found to have whole, active SARS-CoV-2 virus in their stools. Viral shedding in stools lasted up to thirty-three days longer than viral shedding in respiratory secretions. Based on their meta-analysis, the authors concluded that fecal-oral spread was possible with COVID-19.[68]

In a second study, researchers in Hong Kong examined the medical literature, including three Chinese databases, and collected data from fifty-nine patients hospitalized with COVID-19 during February 2020 in Hong Kong. From their analyses, they determined that 17.6 percent of COVID-19 patients had gastrointestinal symptoms. Of all COVID-19 patients, 48 percent had viral RNA in their stool specimens, even those with negative respiratory secretion samples. As with the previous study, these researchers concluded that fecal-oral spread of SARS-CoV-2 was possible.[69]

Neither study proved that fecal oral spread of COVID-19 had occurred, however. In countries with poor sanitation systems, the threat of fecal-oral transmission was a serious concern, as evidenced by an editorial in the *Indian Journal of Gastroenterology*, which recognized the need to avoid open defecation and restricted use of public toilets to reduce potential COVID-19 transmission. Extensive public education campaigns of personal hygiene measures such as handwashing to reduce the spread of SARS-CoV-2 and other deadly microbes would be important.[70] Even public toilets, if available, might pose possible disease transmission risks.[71] But until studies documented the transmission of SARS-CoV-2 by the fecal-oral route, the threat would remain theoretical. (A summary comparison of the R_0, primary mode of spread, airborne transmission, and wastewater detection of SARS-CoV-1, MERS-CoV, and SARS-CoV-2 [COVID-19] is available in appendix 2.)

Fomite Transmission

During the earliest days of the pandemic, images of sanitation workers spraying city streets in Wuhan, China, raised fears globally.[72] In February 2020, the WHO published a report stating that COVID-19 was transmitted by fomites.[73] Fomites are contaminated objects. One month later, a study in the *New England Journal of Medicine* demonstrated that viable virus was detected for up to seventy-two hours on plastic and stainless steel surfaces.[74] The WHO subsequently recommended that people clean and disinfect surfaces to reduce viral spread. Heavy public demand for disinfectants, gloves, and other cleaning supplies led to the products flying off supermarket shelves.[75]

But no reports appeared of proven disease transmission by fomites.

Emanuel Goldman, a professor of microbiology at Rutgers New Jersey Medical School, doubted that fomites spread the virus and wrote an opinion essay in *Lancet Infectious Diseases* in July 2020 stating just that. He argued that the laboratory studies being done used enormous amounts of viruses on surfaces that did not represent real-life situations. Instead, he believed that the risk of transmission by fomites was small and would occur only in cases in which infected individuals sneezed on surfaces followed by others touching the surfaces within one to two hours. He urged a balanced perspective to curb counterproductive excesses.[76]

Subsequent studies supported his position.

For example, one study took repeated samples over a two-week period from household environmental surfaces such as light switches, door handles, toilet handles, refrigerator handles, pillows, nightstands, and other objects in homes in which one family member had laboratory-confirmed COVID-19. The sampling took place from March to April 2020 in ten homes in Salt Lake and Davis Counties, Utah. The researchers found SARS-CoV-2 genetic material (RNA) on 23 (15 percent) out of 150 environmental swabs, most commonly on nightstands, pillows, and light switches used by individuals who had recently tested positive for SARS-CoV-2. Intact, viable SARS-CoV-2 was obtained from only one sample, a nightstand, used by a symptomatic 35-year-old man with a high viral load, early in his course of illness. Of the six out of ten households that had environmental detection of SARS-CoV-2 genetic material, three had more than one secondary transmission from the index case. But none of these households had isolated the sick family member, nor had they routinely cleaned or disinfected any surfaces. The four households that did not have positive environmental samples of SARS-CoV-2 had no secondary transmission cases. The index cases in these households isolated themselves, slept in separate bedrooms, used separate bathrooms, ate alone, and used disinfectants. The authors concluded that the lack of isolation of the index cases more likely contributed to the secondary transmission of COVID-19 than fomites.[77]

This and other studies found surface contamination with viral genetic material or intact, whole coronavirus, but no study proved secondary transmission by fomite. As a result, the WHO concluded that fomite transmission is possible but unlikely.[78] The CDC agreed: "The risk of fomite transmission after a person with COVID-19 has been in an indoor space is minor after 3 days (72 hours), regardless of when it was last cleaned."[79]

Droplet and Airborne Transmission

The debate between the importance of droplet versus airborne transmission of COVID-19 led to a scrutiny of definitions as well as in-depth investigations of exhaled respiratory secretions. Respiratory particles could be defined by how far they travel. Particles that travel short distances, less than 2 meters, are called droplets; whereas particles that travel more than 2 meters are called droplet nuclei or aerosols.[80] For infected respiratory droplets to be inhaled, they must be expelled within close proximity, generally 1.5–2.0 meters, of another individual.

Particles could be defined by size. Coughing, sneezing, talking, or breathing produce globs of saliva, mucus, and water of varying sizes. Bigger globs, greater than 5 to 10 micrometers, are called droplets and fall to the ground. Smaller globs, less than 5 micrometers, are called droplet nuclei, synonymous with "aerosol," and linger in the air like a cloud.[81,82,83]

Investigations of droplet size found the cut-off at 5 micrometers to be inaccurate, however.[84,85] A size threshold of 100 micrometers, not 5 micrometers, might be used, but depending on the force of expulsion and environmental conditions, both droplets and aerosols could fall to the ground or remain suspended. Therefore, the terms "droplet" and "aerosol" were arbitrary and misleading. The term "particles" for both was recommended instead.[86]

Infected individuals could spray virus-laden particles like "tiny cannonballs" to susceptible people. Or they could sneeze virus-laden particles capable of remaining airborne for hours and be inhaled like smoke. In poorly ventilated indoor settings, airborne viruses could travel more than 2 meters, causing superspreading events. Asymptomatic individuals with COVID-19 would be more likely to produce virus-laden particles by talking or breathing, and no one would suspect them of being contagious. Very sick, symptomatic individuals could cough or sneeze particles, and these individuals should be isolated from healthy people. Therefore, public health measures should focus on mandates to wear face masks, which can physically block particles; improving indoor air ventilation; and installing high-efficiency particle air (HEPA) filtration systems, especially in health care and other high-risk indoor environments.[87,88]

In May 2021, the *Lancet* published an editorial outlining ten scientific reasons why SARS-CoV-2 spread by airborne transmission:

1. Substantial numbers of cases due to superspreading events;
2. Long-range transmission between separated people in quarantine hotels;
3. At least a third of spread was from asymptomatic or pre-symptomatic people;
4. Higher transmission rates indoors than outdoors;
5. Hospital-acquired infections despite rigorous contact and droplet precautions;
6. Viable, infectious virus detected in the air up to three hours experimentally and in patients' rooms;
7. Virus identified in hospital building ducts and air filters;
8. Disease spread between animals in separated cages;
9. No evidence to refute airborne transmission;
10. Limited evidence to support other routes of disease transmission.

Based on these findings, the authors argued that there was enough evidence to accept the airborne transmission of SARS-CoV-2 and that it was a scientific error to dismiss this reality because of a lack of direct evidence.[89] The editorial was published over a year after the WHO declared COVID-19 to be a global pandemic, yet health officials were reluctant to accept that the disease spread through the air. Initially, the CDC and WHO discouraged the public from wearing face masks to reduce disease spread.[90] Their policies likely increased the number of people infected by the virus.

Humans

The 1960s Coronavirus Discoveries: B 814, 229 E, OC 43

In the early 1960s, the common cold sparked much research interest. Viruses that caused common colds, such as the rhinovirus, adenovirus, influenza virus, parainfluenza virus, and respiratory syncytial virus, had been discovered already.[1] But others, most notably coronaviruses, had not. While veterinarians had been battling coronavirus outbreaks in animals for thirty years, physicians were largely ignorant of them and of the potential threat they posed to humans.

B 814

Epsom College, a British boarding school for boys ages 12 to 17, was the site of an early common cold study. From June 1960 to June 1961, Dr. E. J. C. Kendall, a general practitioner in Epsom providing medical care in the school's sanitorium, collaborated with Drs. David Tyrrell, a physician virologist and director of the Medical Research Council's Common Cold Research Unit in Salisbury, Wiltshire, and M. L. Bynoe, the medical superintendent of the research unit.[2,3,4] Their goal was to study young people with common cold symptoms to understand respiratory viral illnesses affecting school attendance. They defined common colds as acute infections of upper respiratory tracts with prominent symptoms being nasal blockage and discharge. Of 368 boys boarding in the dormitories, 225 developed common colds during the study period. To collect viruses, the researchers washed the nasal passages of the sick boys with antibiotic-laced saline. The boys blew their nasal fluids, with as much mucus as possible, into petri dishes for further processing. From the washings of 59 boys, the researchers isolated eighteen viruses. One was designated B 814, "from a boy with a typical common cold in 1960."[5]

The laboratory technologies of the 1960s were cumbersome, making the study of common cold viruses time consuming and difficult. Rhinoviruses, adenoviruses, and enteroviruses were cultivated using the standard methodology of the day, namely "tissue cultures," which used proteolytic (protein-destroying) enzymes such as trypsin to break down tissues, allowing cells to be separated from each other. Fetal lung cells were commonly used to grow rhinoviruses and kept alive in glass bottles filled with nutrient-rich broth. The cells would attach to the glass sides of the bottles and replicate. But other respiratory viruses couldn't be cultivated this way.

The researchers developed their own study techniques and supplies, often inspired by the work of other researchers who had described their methodologies in their papers. For example, Tyrrell read a paper published by Bertil Hoorn, a Swedish otolaryngologist who had become intrigued by the causative agents of common colds. Hoorn developed a novel "organ culture" technique using ciliated (lined with tiny hairs) tracheal epithelium tissues that were kept alive with synthetic media.[6] He cut tracheas from human fetuses into 5- to 6-millimeter square pieces using the cartilage as scaffolding. He knew the tissues were viable because the cilia would beat. When exposed to viruses such as influenza and adenoviruses, the cilia stopped beating. The herpes simplex virus destroyed the entire epithelium.[7]

Tyrrell decided to try Hoorn's technique and cultivated B 814 in organ cultures using the tracheas of 14- to 22-week-old human embryos obtained from hysterotomies. The experiments worked. B 814 could be cultivated and proved to be sensitive to ether, meaning that it had a lipid coat. He also showed that it was an RNA virus by putting a DNA inhibitor into the organ culture. Based on B 814's unique characteristics, Tyrrell and Bynoe concluded that it was a new virus.[8]

229 E

Meanwhile, in the United States, Drs. Dorothy Hamre, a microbiologist, and John Procknow, an infectious disease physician, both with the Preventive Medicine Section of the University of Chicago's Department of

Medicine, were conducting a five-year surveillance study of mild upper respiratory tract infections in medical students. In the winter of 1962, they isolated a novel virus from four students with colds and from one healthy student. They determined it was an RNA virus and called it 229 E.[9,10,11]

In 1966, Dr. June Almeida, an electron microscopist and virologist, working at the St. Thomas's Hospital Medical School Department of Medical Microbiology in London, collaborated with Tyrrell to visualize B 814 and 229 E, sent courtesy of Hamre and Procknow. Using the electron microscope, they noted that the two viruses appeared indistinguishable and were morphologically identical with the avian infectious bronchitis virus.[12] Their findings prompted them and six colleagues to name the new group of viruses coronaviruses.[13] B 814 was the first coronavirus identified in humans, and 229 E was the second.[14,15,16]

Tyrrell's Common Cold Research Unit proved Koch's postulates that coronaviruses caused common colds. Using 229 E and B 814 in two separate batches, they dripped virus-laden fluids of varying doses into the nostrils of male and female volunteers, ages 18 to 50. The volunteers were recruited through advertisements such as "Free 10 Day Autumn or Winter Break: You May Not Win a Nobel Prize, but You Could Help Find a Cure for the Common Cold."[17] Thirteen (50 percent) out of twenty-six volunteers and thirty-four (45 percent) out of seventy-five developed colds from 229 E and B 814, respectively. The volunteers who received the highest doses of viruses had the highest rates of illnesses. Cold symptoms ranged from mild to severe, but no one developed pneumonia.[18]

OC 43

At the National Institutes of Health (NIH) in Bethesda, Maryland, physician Dr. Kenneth McIntosh and his colleagues at the Laboratory of Infectious Diseases were conducting surveillance studies of volunteer NIH employees with acute upper respiratory tract infections. They used the same organ culture (OC) techniques developed by Hoorn, Tyrrell, and Bynoe to isolate eight viruses from twenty-three volunteers. Upon electron microscopy, six of the viruses appeared identical to 229 E and to avian infectious bronchitis virus (IBV).[19] The "IBV-like" virus from the volunteers also appeared identical to the mouse hepatitis virus

(MHV). Two of the six viral strains, designated OC 38 and OC 43, grew in suckling (baby) mouse brains, causing signs of encephalitis: lethargy, rigidity, and generalized tremors.[20] This finding was important because it enabled McIntosh to grow large amounts of OC 38 and OC 43 in mice without needing organ cultures, which were difficult to perform and extremely expensive. Subsequent immunological tests in mice revealed that OC 38 and OC 43 were the same virus.[21]

According to McIntosh, nobody else wanted to work with coronaviruses because they only caused common colds and were a total pain to work with.[22] As a result, thirty-five years passed between OC 43 and the next coronavirus to appear in humans: SARS-CoV-1.[23]

The 2000s Coronavirus Discoveries: SARS-CoV-1, CoV-NL63, CoV-HKU-1, MERS-CoV, SARS-CoV-2

SARS-CoV-1

The Program for Monitoring Emerging Diseases (ProMED) was founded in 1994 by Drs. Stephen S. Morse, an epidemiologist with the Columbia University School of Public Health, and Barbara Hatch Rosenberg, a bioweapons expert at the Federation of American Scientists, to improve the rapid detection of emerging disease outbreaks and to disseminate information in real time to a global audience.[24] ProMED worked. It provided an early warning of an emerging disease: SARS-CoV-1.

On February 10, 2003, the first inkling of an epidemic appeared in a ProMED post by Dr. Stephen O. Cunnion, an infectious disease outbreak consultant with a military background. He had received a personal email from a former neighbor in the US Navy, a schoolteacher, who was participating in an international teachers chat room when a Chinese member told her about people dying in the streets and hospital lockdowns.[25]

Dr. Cunnion wrote in ProMED (italics in original):

> *This morning I received this e-mail and then searched your archives and found nothing that pertained to it. Does anyone know anything about this problem?*
>
> *"Have you heard of an epidemic in Guangzhou? An acquaintance of mine from a teacher's chat room lives there and reports that the hospitals there have been closed and people are dying."*[26]

A more ominous ProMED posting appeared the next day:

> *An unidentified pneumonia virus has killed 5 people and left [300] hospitalized in southern China, while rumors of a surging death toll prompted frightened residents to stock up on antibiotics . . . [it] was first reported in November [2002] in four cities in Quangdong [province] . . . "We did not take it seriously at the beginning," said the [provincial disease-control] official.*[27]

The epidemic of atypical pneumonia began November 16, 2002, in the city of Foshan in China's Guangdong Province. At that time, the Guangdong Province had a population of 85.2 million people. An epidemiologic investigation of the initial five months of the epidemic found a total of 1,454 SARS cases and 55 deaths in the Guangdong Province with case-fatality rates of 3.8 percent for all ages and 12.7 percent for people older than 65 years. Nine of the 23 earliest cases, infected on or before January 2003, were food handlers: people who handled, killed, or sold food animals as well as those who prepared or served food. Seven were chefs working in restaurants where animals were slaughtered on the premises. The two others meeting the case definition for "food handler" were a snake seller and a market produce buyer. Notably, none were farmers raising livestock or poultry.[28]

The earliest known patient, identified by retrospective case searching, lived in Foshan city with his wife and four children and had prepared food involving chickens, domestic cats, and snakes. He infected his wife, his 50-year-old aunt, his aunt's 50-year-old husband, and his aunt's 22-year-old daughter. None of his children got sick. The second-earliest known patient lived in Heyuan but worked as a chef in Shenzhen. His work did not involve killing animals. Many of the earliest cases infected family members and health care workers. The study authors speculated that the epidemic began either by a single point source or by several point sources in the Pearl River basin.[29]

A different epidemiologic study also conducted in the Guangdong Province tested 792 people for immunoglobulin G (IgG) antibodies to SARS-CoV-1.[30] Like microscopic missiles, IgG antibodies are the most common antibodies in the blood and, if elevated, can indicate infection, among other things. The subjects worked either in animal markets (508), health care or public health (200), or were healthy controls (84). Of the 72 subjects who tested positive, 66 worked in animal markets.

Those who traded primarily masked palm civets had the highest positivity rates, almost 73 percent, but none had experienced atypical pneumonias, suggesting asymptomatic infections with either SARS-CoV-1 or with antigenically related viruses.[31]

The Chinese government, which had not notified the international community about the growing outbreak for four months, reassured the world that the disease was under control and subsequently hindered further investigations.[32] Nevertheless, on March 12, 2003, the World Health Organization (WHO) issued a global alert for the atypical pneumonia. Two days later, the Centers for Disease Control and Prevention (CDC) activated its Emergency Operations Center, and on March 21, a *Morbidity and Mortality Weekly Report* summarized the clinical description of SARS patients, most of whom were healthy adults between the ages of 25 and 70.

After an incubation period between two and ten days, the illness would begin with fevers, chills, and rigors, sometimes accompanied by malaise, myalgias, and headaches. Asymptomatic and mild infections were uncommon and did not appear infectious.[33] Some patients reported diarrhea. After the prodrome, the earliest signs or symptoms of illness, the lower respiratory phase began with a dry, nonproductive cough and dyspnea that could progress to hypoxemia (low blood oxygen levels).[34] Twenty to 30 percent of SARS patients required intensive care, including mechanical ventilation.[35]

On May 15, 2003, the *New England Journal of Medicine* dedicated virtually its entire issue to SARS.[36] First, it published an in-depth report of a SARS outbreak in the Prince of Wales Hospital in Hong Kong involving 138 cases, 50 percent of whom were health care workers. Chest radiographs showed air-space consolidation indicating pneumonia with predominantly peripheral zones involved. Progression of pulmonary infiltrates indicated worsening pneumonias. Thirty-two (23 percent) of the patients developed respiratory failure, and of these, 19 (13.8 percent) required mechanical ventilation. Almost 67 percent of the patients who needed intensive care were male, with mean ages around 50 years, but they did not necessarily have preexisting conditions. Five patients died.[37]

In addition, the journal published two reports: one from the CDC and the other from Europe, involving collaborating researchers in Germany,

France, and the Netherlands. The groups obtained clinical specimens from SARS patients and used traditional cell cultures, electron microscopy, immunological and serological tests, and novel molecular analyses including reverse-transcriptase polymerase chain reactions (RT-PCR) to identify the novel coronavirus.[38]

In the decades after the discovery of OC 43, the development of polymerase chain reaction (PCR) tests revolutionized virology. In 1983, Dr. Kary Mullis, a biochemist working in the human genetics department at the Cetus Corporation, an early biotechnology company based in Northern California, developed a methodology that essentially made billions of "DNA photocopies," facilitating detailed genetic and molecular analyses.[39] The work was initially done to aid in diagnosis of sickle cell anemia, but it soon became apparent that it had widespread applications in medicine and biomedical research.[40]

A real-time RT-PCR assay was developed with the goal of performing rapid diagnostic testing during the SARS pandemic.[41] But in 2003 the technology remained expensive and labor intensive, required expertise, and was vulnerable to contamination that could lead to false-positive results. As a result, the gold standard for confirming a SARS diagnosis was a whole-virus immunoassay—either an immunofluorescence or an enzyme-linked immunosorbent assay (ELISA) test. The ELISA test typically used two different antibodies to bind to different regions of the targeted virus. The antibodies were obtained from rabbits and guinea pigs that had been immunized with SARS nucleocapsid proteins.[42,43]

Analyzing the SARS virus involved whole genome sequencing (WGS)—the identification of every single nucleotide in the viral genome. WGS is analogous to identifying every single letter in an entire book. Developed in 1976, WGS set the stage for understanding viral evolution by tracking entire viral genomes over time.[44] In the ensuing decades, automation and other technological advances allowed scientists to identify the emergence of new variants during epidemics.[45]

For example, the Chinese SARS Molecular Epidemiology Consortium, a large team of researchers and epidemiologists from academia and government agencies in the cities of Hong Kong, Guangzhou, Shanghai, and Wuhan, used WGS to track the earliest cases and molecular evolution of the SARS virus over the course of the epidemic. In the beginning of the epidemic, they found eleven seemingly independent index cases, seven of which had had contact with wild animals. These cases appeared in the Pearl River Delta region, a major manufacturing hub and one of the most densely populated regions in the world, in the Guangdong Province, known for its culinary culture of consuming exotic animals.[46]

They determined that two major viral genotypes predominated during this initial phase of the epidemic. The first genotype contained a 29-nucleotide sequence that was virtually identical to those isolated from wild animals for sale and from the humans working in the live animal market in Shenzhen.[47] The second genotype had an 82-nucleotide sequence deletion that was found in the earliest cases in the city of Zhongshan and was identical to coronaviruses isolated from farmed civet cats in the Hubei Province. The consortium conclusion: there was strong evidence for an animal origin of the virus.[48]

Relatively few variants appeared during the epidemic. A variant emerged in Hong Kong toward the end of the epidemic that was traced to an 85-year-old hospitalized woman, in whom SARS had not been initially suspected, but who had subsequently tested positive. She infected seven other inpatients and two health care workers. The SARS-CoV-1 variant first appeared in mid-April 2003 and infected three generations, but it did not appear to be more infectious or virulent than the initial viral strains.[49]

Some survivors suffered from long-term mental and physical disabilities. Most studied were health care workers. Depression, post-traumatic stress disorder, and panic disorder were some of the psychiatric sequelae. Long-term physical sequelae included reduced pulmonary function and osteonecrosis of the hip joints from steroid therapy given during the acute phase of the SARS infection.[50]

SARS was contained through the international collaboration and cooperation of physicians, scientists, and epidemiologists who identified the coronavirus, developed reliable diagnostic tests to track it, and implemented containment strategies to stop its spread.[51] Epidemiologists confirmed that health care workers were at heightened risk.[52] Since only people who were sick were infectious, public health measures like quarantine and isolation worked.[53] On July 5, 2003, after eight months, the WHO declared the SARS epidemic had been contained. The coronavirus had spread to twenty-nine countries and regions, infected over 8,000 people, and killed 774.[54]

CoV-NL63

The year after SARS ended, Dr. Ron Fouchier, a virologist at the Erasmus Medical Center in Rotterdam, the Netherlands, and his colleagues published a report of a new coronavirus that had caused pneumonia in an 8-month-old boy, hospitalized sixteen years earlier. The virologists had been testing archived clinical specimens from cases in which the causative agent had not been identified, an effort reminiscent of the search for novel cold viruses.

By this time, laboratory supplies had become big business. Research tools were being increasingly sold as sophisticated "kits" by multinational corporations. For example, advances in biotechnology had allowed Fouchier to grow the virus in specialized cells, isolate its RNA using a "High Pure RNA Isolation kit" made by Roche Diagnostics (a Swiss multinational health care company), sequence its entire genome using RT-PCR, create a complementary DNA copy of the virus's RNA, and develop a phylogenetic tree to determine its lineage. Genetic analysis revealed that the new coronavirus was closely related to the porcine epidemic diarrhea virus and to CoV-229E. Fouchier and his colleagues named the new virus HCoV-NL. Testing of additional archived cases over a two-year period (2000–2002) found 4 out of 139 samples positive for HCoV-NL in hospitalized children with chronic underlying medical conditions and severe acute respiratory tract infections.[55]

Around the same time that Fouchier was virus hunting, a Dutch team at the University of Amsterdam isolated a coronavirus from a hospitalized 7-month-old child with bronchiolitis and conjunctivitis. A nasopharyngeal aspirate was taken from the baby, and the sample was labeled NL63. Diagnostic tests for the usual viral suspects such as respiratory syncytial virus, adenovirus, influenza and parainfluenza viruses, enterovirus, and others were negative. The researchers theorized that the causative agent might be a new virus, so they developed their own novel laboratory test called VIDISCA that allowed them to amplify the genome of any pathogenic virus, known or unknown.[56] Processing the child's aspirate using VIDISCA yielded fragments of a coronavirus. They concluded that it was a new coronavirus, sequenced its entire genome, and found it to be closely related to CoV-229E. They named it CoV-NL63. After screening archived respiratory specimens from outpatient clinic visits the previous year, the researchers found seven additional patients infected with CoV-NL63.[57]

Both Dutch papers were published independently in April 2004.

Since 2004, CoV-NL63 has been found to infect both upper and lower respiratory tracts and has been detected in approximately 1 to 9 percent of respiratory tract samples in at least eighteen countries. Upper respiratory tract infections are usually mild, with symptoms involving cough, sore throat, rhinitis, and fever. Lower respiratory tract infections due to CoV-NL63 can be severe, particularly in children, causing bronchiolitis (inflammation and congestion of small airways) or croup (a barking cough due to inflammation and swelling around the larynx, trachea, and bronchi), and requiring hospitalization.[58]

The virus has been associated with Kawasaki disease, a systemic vasculitis (inflammation of blood vessels) in children. Symptoms of Kawasaki disease include fever for at least five days, erythema of the mouth or pharynx, stomatitis or strawberry tongue, bilateral conjunctivitis, edema or erythema of hands or feet, and cervical lymphadenopathy (swelling of lymph nodes). Individuals must have at least three of these symptoms to meet the diagnostic criteria. There are no laboratory tests that confirm the diagnosis of Kawasaki disease.[59]

The coronavirus has zoonotic origins in bats. A surveillance study of bat coronaviruses in Kenya from 2007 to 2010 found genetic evidence that HCoV-NL63 developed when its genetic precursors in *Triaenops* bats combined with 229 E–like viruses in *Hipposideros* bats.[60]

CoV-HKU1

In Hong Kong, Dr. Patrick C. Y. Woo, a physician-microbiologist at the State Key Laboratory of Emerging Infectious Diseases at the University of Hong Kong, and his colleagues in collaborating academic medical centers found another new coronavirus. Their index patient was a 71-year-old Chinese man with acute pneumonia. He had a forty-year history of tuberculosis, was a chronic smoker, had chronic obstructive lung disease, and had a chronic *Pseudomonas aeruginosa* infection in his airways. In January 2004, he was hospitalized three days after returning from a trip to Shenzhen, China, but had tested negative for SARS.

But viral RNA was isolated from his respiratory secretions, his urine, and his fecal specimens. RT-PCR identified a coronavirus. The researchers sequenced its genome, compared it to other known coronaviruses, and found that it was distantly related to murine hepatitis virus, CoV-OC43, and bovine coronavirus. They named it CoV-HKU1. It appeared to be rare. Out of four hundred clinical specimens (nasopharyngeal aspirates) collected the preceding year from patients with respiratory illnesses, only one was positive for CoV-HKU1. It was from a 35-year-old previously healthy woman with pneumonia of unknown etiology. She had been sick in March 2003, during the SARS epidemic, ten months earlier than the 71-year-old man. She had tested negative for SARS. There was no relationship between the two patients.[61]

After its discovery, CoV-HKU1 had been reported in patients with pneumonia and other acute respiratory illnesses in Australia, France, the United States, Italy, Switzerland, Thailand, and Brazil.[62,63] Some patients were also found to be positive with other coronaviruses, namely CoV-OC43, CoV-229E, and CoV-NL63. CoV-HKU1 has been found in rodents (see https://www.ncbi.nlm.nih.gov/pmc/articles/PMC7098031/ for more information).

MERS-CoV

Like SARS, the first hint of MERS appeared in a ProMED posting. The date was September 20, 2012. The subject line was "Novel coronavirus—Saudi Arabia: human isolate." Dr. Ali Mohamed Zaki, a professor of microbiology at Dr. Soliman Fakeeh Hospital in Jeddah, Saudi Arabia, wrote the post. The index patient, a 60-year-old male, had been hospitalized with pneumonia. Clinical specimens had tested negative for influenza A and B, parainfluenza, enterovirus, and adenovirus but had tested positive for coronavirus.[64] He sent the specimens to Dr. Ron Fouchier in the Netherlands to confirm that it was indeed a new coronavirus.[65]

On November 8, 2012, they coauthored a report in the *New England Journal of Medicine* describing the index patient and the process of identifying the novel coronavirus. The 60-year-old Saudi male had experienced seven days of fever, cough, and shortness of breath. He did not smoke, was taking no medications, and had no history of chronic illnesses. He was hospitalized on June 13, 2012, with a chest X-ray showing bilateral pneumonia. On the day after hospital admission, he was transferred to the intensive care unit for mechanical ventilation.

Despite intensive medical care, he deteriorated. Broad-spectrum antibiotics were ineffective. His kidney function began to fail. His liver enzymes increased. On the eleventh day, he died from respiratory and renal failure.[66]

The search for the offending pathogen included collecting blood and sputum samples and processing them using advanced technologies. The specimens were purified to remove unwanted contaminating materials using research tools such as the Ribo-Zero rRNA Removal Kit by Epicenter, a brand of Illumina, a US-based corporation that develops, markets, and supplies technologies to do genetic sequencing.[67] Viral RNA fragments were extracted from cell cultures using the High Pure Viral Nucleic Acid Kit made by Roche. The fragments were sequenced using a BigDye Terminator v3.1 Cycle Sequencing Kit and a 3130XL Genetic Analyzer made by Applied Biosystems, a brand under the Life Technologies brand of Thermo Fisher Scientific Corporation, a multibillion-dollar corporation that supplies life sciences researchers with diagnostic and other laboratory technologies.[68] A library of the virus's RNA fragments was made using Roche's GS FLX Titanium Rapid Library Preparation platform. The researchers created a phylogenetic tree to assess the lineage of the new coronavirus. It was a beta coronavirus like SARS-CoV-1, CoV-OC43, and CoV-HKU1 but more closely related to two bat coronaviruses. They named the virus HCoV-EMC for the Erasmus Medical Center where Dr. Fouchier worked.[69]

After the initial patient, more patients appeared with severe pneumonia, multi-organ failure, and epidemiologic links to countries in the Middle East, including Jordan, Qatar, United Arab Emirates, and especially Saudi Arabia. Given the geographic connections, the Coronavirus Working Group of the International Committee on Taxonomy of Viruses renamed the new coronavirus Middle East respiratory syndrome coronavirus (MERS-CoV).[70] Genomic analysis suggested that the MERS-CoV causing the illnesses had diverged from a common ancestor as early as mid-2011.[71]

A serological survey of over 10,000 healthy adults in all thirteen provinces of Saudi Arabia from 2012 to 2013 found MERS-CoV antibodies, indicating previous infections, in 0.15 percent of the samples. Most of the cases that had tested positive were male and had a mean age of 43.5 years, about a decade younger than the hospitalized MERS patients. The study also collected serum samples from 140 slaughterhouse workers and 87 camel shepherds with occupational exposures to camels. The camel shepherds' MERS-CoV antibody seroprevalence was about fifteen times higher than in the general population, and in slaughterhouse workers, it was about twenty-three times higher. The study suggested that MERS-CoV had been circulating in Saudi Arabia long before first being identified in hospitalized patients.[72] In other words, asymptomatic cases of MERS existed.

After three years, a clearer picture of MERS emerged. Infections ranged from asymptomatic to mild upper respiratory illnesses to severe pneumonias and death. In children, the disease was rare. Patients typically presented to hospitals with fever, cough, shortness of breath, and radiological images of pneumonia. They could also present with headaches, fatigue, muscle aches, vomiting, and diarrhea. Severe MERS infections could include renal failure, acute liver injury, coagulopathy, and cardiac dysfunction. Diagnosing MERS-CoV RNA involved real-time PCR. Respiratory samples could remain positive an average of fifteen to sixteen days, with one study reporting positivity up to thirty-four days.[73] The virus could be also detected in urine and stool thirteen to sixteen days after disease onset, respectively. Treatment was primarily supportive. The overall mortality was approximately 35.6 percent.[74]

From 2012 to 2021, a total of 2,567 laboratory-confirmed cases of MERS and 884 deaths were reported in twelve countries. Eighty-four percent of the cases and 91 percent of the deaths occurred in Saudi Arabia. In 2021, only one case, a male Saudi national, was reported. There were no secondary cases.[75] MERS, like SARS, caused long-term pulmonary function abnormalities, reduced exercise capacities, and psychological impairments in disease survivors.[76]

Human-Porcine Deltacoronavirus (Hu-PDCoV)

Between May 2014 and December 2015, three Haitian schoolchildren, ages 6 to 7 years, with symptoms including fevers, coughing, and abdominal pain, were found to be infected with porcine deltacoronavirus. It was the first time that deltacoronavirus infections had been identified in humans. Whole genome sequencing studies revealed that the viral strain in two of the children was 99.97 percent similar and closely related to a pig strain detected one year earlier in Tianjin, China. The third child, who attended a different school, had a variant closely related to a pig strain identified in Arkansas in 2015.

After culling its pig populations in the 1980s to eliminate African swine fever, Haiti began importing pigs from China, Europe, and North America. The Hu-PDCoV strains likely had been circulating in Haitian pigs for several years before the spillover event, but there was no surveillance system to monitor viral infections in the farm animals to confirm this assumption. The children were very poor and lived in a rural, mountainous region of the country. All three recovered from their infections.[77]

SARS-CoV-2

As with SARS and MERS, information about the COVID-19 pandemic began with ProMED. On December 30, 2019, Dr. Marjorie Pollack, a physician-epidemiologist and ProMED moderator based in New York City, received an email from a colleague who tracked Weibo, a social media platform in China, about posts describing seriously ill patients in Wuhan, China. Pollack contacted ProMED colleagues for additional information. Several hours later, HealthMap, an artificial intelligence (AI) system at Boston Children's Hospital, issued an alert from a translated report from *Sina Finance*, a Chinese technology company based in Beijing, describing clusters of unidentified pneumonia cases in Wuhan.[78]

After accumulating the information, Dr. Pollack issued a ProMED alert titled "Undiagnosed pneumonia—China (Hubei): Request for Information." The post, based on the AI translation, included the following:

> On the evening of [30 Dec 2019], an "urgent notice on the treatment of pneumonia of unknown cause" was issued . . . by . . . the Medical Administration of Wuhan Municipal Health Committee. On the morning of [31 Dec 2019] China Business News reporter called the official hotline of Wuhan Municipal Health and Health Committee 12320 and learned that the content of the document is true . . . Another . . . emergency notification, entitled "City Health and Health Commission's Report on Reporting the Treatment of Unknown Cause of Pneumonia," is also true . . . the South China Seafood Market in our city has seen patients with pneumonia of unknown cause . . . The so-called unexplained pneumonia cases

> refer to the following 4 cases of pneumonia that cannot be diagnosed . . . : fever (greater than or equal to 38C); imaging characteristics of pneumonia or acute respiratory distress syndrome; reduced or normal white blood cells . . . The number of lymphocytes was reduced. After treatment with antibiotics for 3 to 5 days, the condition did not improve significantly. It is understood that the 1st patient with unexplained pneumonia that appeared in Wuhan . . . came from Wuhan South China Seafood Market . . . Wuhan CDC went to treatment hospital to collect patient samples . . . Wuhan has the best virus research institution in the country, and the virus detection results will be released to the public as soon as they are found.[79]

The next day, Chinese authorities notified the WHO China Country Office of pneumonia cases of unknown etiology in Wuhan City, Hubei Province. On January 3, 2020, they reported a total of forty-four patients with undiagnosed pneumonia: eleven severely ill and thirty-three stable. The patients' chest radiographs showed bilateral pneumonia. All were in isolation. Some of the patients, according to the Chinese authorities, were vendors or operating dealers in the Huanan seafood market.[80]

On February 15, 2020, Chinese clinicians and researchers published two separate reports in the British journal *Lancet* describing the clinical and epidemiological characteristics of the patients hospitalized with the novel coronavirus in Wuhan. In the first report, forty-one patients with pneumonia were admitted from December 16, 2019, to January 2, 2020, to isolation wards with airborne precautions in Jin Yin-tan Hospital in Wuhan, an infectious disease specialty hospital for adults 14 years and older. The diagnoses were confirmed using real-time RT-PCR with AgPath-ID One-Step RT-PCR Reagent, made by Thermo Fisher Scientific. Real-time PCR, also known as quantitative PCR (qPCR), was the gold standard for detection and quantification of viral nucleic acid targets. The test was simple, fast, and efficient.

The first reported patient had developed symptoms on December 1, 2019. There were no epidemiologic connections between him and any of the other patients, nor did he infect any of his family members. He had no exposure to the Huanan market, but twenty-seven (66 percent) of the other patients did, including the first fatal case. The median age of the patients was 49 years. Seventy-three percent of the patients were male, and 32 percent had comorbidities. All had pneumonia with abnormal chest CT scans, and virtually all had bilateral lung involvement. Thirteen patients required intensive care, and six died. In their conclusions, the study authors wrote, "We are concerned that 2019-nCoV could have acquired the ability for efficient human transmission. Airborne precautions, such as fit-tested N95 respirator, and other personal protective equipment are strongly recommended."[81]

In the second report, ninety-nine patients with the novel coronavirus were transferred from other Wuhan-area hospitals to Jin Yin-tan Hospital, the same hospital as the first report, between January 1 and January 20, 2020. Two of the patients were a married couple, and almost half of the cases were clustered, meaning that they were epidemiologically related. Forty-seven had long-term exposures at the Huanan market, including employment histories as market managers or salesmen. None were explicitly reported to be chefs or individuals employed to butcher animals. The mean age was 55.5 years, and 68 percent were males. Fifty-one percent had chronic medical illnesses, such as cardiovascular or endocrine diseases. Twenty-three percent were admitted for intensive care, and 11 percent died.[82]

A scientific debate ensued as to whether earlier COVID-19 cases might have gone undetected. One study used viral genome data to conduct epidemiological simulations and determined that SARS-CoV-2 was likely circulating in the Hubei Province at low levels as early as October but more likely sometime in November 2019.[83] Evidence to support that assertion came from Chinese government records suggesting that the earliest patient was a 55-year-old male from Hubei Province who contracted the virus on November 17, 2019. But earlier cases could not be ruled out.[84]

By early 2020, Chinese physicians had conducted many serological surveys, testing thousands of people, to assess how common antibodies to SARS-CoV-2 were in the general population, both inside and outside Wuhan. Unlike SARS and MERS, however, none of the studies explicitly included occupational exposures and antibody seroprevalence rates. Such studies would have been important because they might have revealed whether animal workers in the Huanan market had higher rates of SARS-CoV-2 antibodies than the general population, supporting the natural spillover hypothesis. The studies showed that the highest seroprevalence rates were in Wuhan compared to other parts of China, which would be expected since the pandemic began there. Older people, particularly women, were more likely to test positive for SARS-CoV-2 antibodies. Antibody titers were lower in individuals who had asymptomatic versus symptomatic infections.[85,86,87,88,89,90]

In the wake of the SARS epidemic of 2002–2003, China had established a Viral Pneumonia of Unknown Etiology (VPUE) surveillance system to pick up the earliest cases of emerging diseases. But there were problems with the system, most notably a lack of awareness by most clinicians, who were not reporting to it. Despite these problems, one study concluded that there was strong evidence of a live animal market origin of the pandemic because most of the earliest cases lived near the Huanan market. However, the study author acknowledged that no live mammal in the market had been screened for SARS-CoV-2-related viruses and that the market had been closed and disinfected on January 1, 2020, prohibiting conclusive findings.[91]

Another study examined the clinical and epidemiological features of 326 patients in Shanghai between January and February 2020. The clinicians classified the patients into four infection categories: asymptomatic, mild, severe, and critical. Almost 90 percent of the patients had mild infections with fever and radiological signs of pneumonia. The severe patients (4 percent) were hospitalized, and the critical patients (5 percent) required mechanical ventilation or extracorporeal membrane oxygenation. Host factors determining infection severity included comorbidities such as hypertension and diabetes, male gender, lymphocytopenia, and cytokine levels. Low lymphocyte counts and high levels of proinflammatory cytokines, particularly interleukins 6 and 8, predicted critical disease.[92]

The clinical manifestations of COVID-19 were vast. A *Science* article published in April 2020 stated, "The virus acts like no pathogen humanity has ever seen." Although the virus typically entered directly or indirectly through the respiratory tract, it affected many organ systems, including the hematologic, cardiovascular, renal, gastrointestinal, hepatobiliary, endocrine, dermatologic, neurologic, and ophthalmologic.[93] One medical review article described the disease's impact on humans as "the four horsemen of a viral Apocalypse."[94] It damaged cells directly, caused thrombo-inflammation, and dysregulated the renin-angiotensin-aldosterone and immune systems.[95]

A brief comparison of the clinical manifestations of SARS, MERS, and COVID-19 are listed in appendix 3a. Most notably, COVID-19 caused loss of taste and smell in many individuals. Loss of smell might be due to damage to the cilia and olfactory epithelium rather than to direct damage to the olfactory neurons.[96] In addition, only COVID-19 had reports of dermatologic manifestations.[97]

Unlike SARS and MERS, which infected thousands, not millions, of people, COVID-19 resulted in many chronic conditions, with myriad names such as "long COVID-19," "post-acute COVID-19," "chronic COVID-19," "persistent COVID-19 symptoms," and "long-haulers." In 2020, the CDC defined chronic COVID-19 as symptoms, signs, or abnormal clinical manifestations persisting two weeks or longer after an RT-PCR confirmed SARS-CoV-2 diagnosis and not returning to a usual state of health.[98] Numerous studies have been published analyzing chronic COVID-19.

One retrospective cohort study conducted in the Netherlands screened electronic health records of 81 million people, including almost 275,000 COVID-19 survivors. Of these survivors, over half were female with a mean age of 46 years and had at least one long-COVID feature within six months of acute illness. The most common symptoms were anxiety and depression, abnormal breathing, abdominal symptoms, fatigue, chest and throat pain, pain, headaches, cognitive symptoms, myalgias, and combinations of these symptoms. The risk of developing chronic symptoms was higher in patients who had more severe disease. Notably, infection from SARS-CoV-2 had a higher risk of producing chronic symptoms compared to influenza, another respiratory virus.[99]

A consortium of researchers in the United States and Mexico screened 18,251 publications, winnowing them down to 15 studies (8 from Europe and the United Kingdom, 3 from the United States, and 1 each from Australia, China, Egypt, and Mexico) that met the criteria of analyzing the long-term complications of COVID-19. The studies were surveys with samples sizes ranging from 102 to 44,779 patients and involving adult patients ages 17 to 87 years. Six studies focused solely on previously hospitalized patients with severe disease. The rest included those with mild or moderate disease. There was no mention of control groups that might also suffer from chronic conditions. Nevertheless, the studies identified fifty-five long-term symptoms, including fatigue, headache, attention disorder, and dyspnea associated with long-term effects of COVID-19. Eighty percent of the patients had at least one symptom, and 34 percent had abnormal chest radiographs or CT scans.[100]

While individuals with severe or critical infections have been most at risk for developing chronic COVID-19 symptoms, those with mild infections also have been vulnerable. As new information about long-term sufferers of COVID-19 accumulates, clinicians have further subdivided the condition into "post-acute COVID-19" and "chronic COVID-19," defining them as symptoms exceeding three weeks and twelve weeks from initial illness, respectively, to better understand the entities they were dealing with.[101]

Investigations into the causes of chronic COVID symptoms suggest myriad possible etiologies, including persistence of the virus or viral RNA or proteins in bodily tissues, minute blood clots, and immune systems abnormalities. Combinations of these factors could also contribute to chronic symptoms.[102] In September 2021, the NIH awarded almost $470 million to researchers at over thirty institutions in its Researching COVID to Enhance Recovery (RECOVER) program to study long COVID.[103]

SARS-CoV-2 is the third major coronavirus to appear in humans in the twenty-first century. (Appendix 3b summarizes human coronaviruses discovered as of 2019.) It has infected hundreds of millions of people and has killed millions. As this book is being written, it continues to mutate into new variants, causing new waves of infections and deaths.

CHAPTER FOUR

Molecular Biology of Coronaviruses

Background

Viruses are clever parasites. During the transcription process, which takes place in a eukaryotic cell's nucleus, like a vault guarding a precious family heirloom, the DNA double helix unzips and exposes its nucleotides (A, C, G, and T), which are copied in the form of messenger RNA.[1] The messenger RNA is subsequently transported to the cell's cytoplasm, akin to a factory floor, where it serves as a blueprint during the translation process to make new proteins. Viruses take advantage of the translation process by commandeering it with their own messenger RNA to reproduce themselves.

In 1971, David Baltimore, a molecular biologist specializing in animal viruses and recently recruited as an associate professor of microbiology in the Department of Biology at the Massachusetts Institute of Technology, developed a classification system of animal viruses based on their genomes.[2]

He divided viruses into six classes:

Class I double-stranded DNA viruses,
Class II single-stranded DNA viruses,
Class III double-stranded RNA viruses,
Class IV positive-sense + single-stranded RNA viruses,
Class V negative-sense—single-stranded RNA viruses, and
Class VI reverse-transcription positive-sense + single-stranded RNA viruses.[3]

Later, a seventh class was added:

Class VII reverse-transcription double-stranded DNA viruses.[4]

Baltimore astutely observed that regardless of the type of genome a virus possessed, after inserting itself into a cell's cytoplasm, the virus needed to convert its genome into messenger RNA before it could harness the cell's machinery for its own uses. For example, double-stranded DNA viruses typically infect bacteria and archaea, a third branch of unicellular life containing methanogenic bacteria, taking advantage of the fact that these microbes don't store their DNA in nuclei the way eukaryotic animal and plant cells do.[5,6] This arrangement allows DNA viruses either to take advantage of the microbes' transcription and translation capabilities or to encode their own.[7] (For a brief description of how scientists discovered messenger RNA and its potential use in medicine and public health, please read appendix 4.)

Over 50 and 90 percent of the known viruses that infect animals and plants, respectively, are single-stranded RNA viruses. Of the animal viruses, almost 28 percent are positive-sense single-stranded RNA viruses—the same configuration as messenger RNA.[8] Coronaviruses are part of this group. They are Class IV single-stranded RNA viruses with linear, non-segmented genomes. In contrast, influenza viruses are Class V negative-sense single-stranded RNA viruses with segmented genomes.[9]

As a group, RNA viruses have notoriously high mutation rates, which are correlated with enhanced evolvability and virulence. High viral mutation rates combined with short generation times and the enormous numbers of viral particles produced during replication help drive viral evolution. Multiplicity of infection (MOI) is the ratio of infectious agents to target cells and is used to explain viral evolution. For example, if the number of infectious agents (e.g., viruses) outnumber the target host cells, then the MOI is greater than one, and the cells can be infected by many viruses. Theoretically, this scenario leads to increased rates of genetic recombination and reassortment. An MOI of less than one, meaning that the target host cells outnumber the viruses, reduces this likelihood.[10] Of course, viral evolution is far more complicated than this, and efforts to understand it are of great scientific and medical interest.

Viral mutations can be beneficial, neutral, or deleterious.[11] RNA virus genomes range in size from 2 to 32 kilobases. Smaller genome sizes appear to correlate with higher mutation rates.[12,13] Of the known RNA viruses, coronaviruses possess some of the largest genomes, perhaps because they have genes that encode for proofreading enzymes, improving replication fidelity.[14,15]

Notably, the original SARS-CoV-2 strain mutated during the COVID-19 pandemic into multiple variants of concern (VOCs), including the highly transmissible Delta and Omicron variants. In general, novel VOC with traits that improve "fitness" facilitate viral transmission and reproduction. These traits manifest as higher basic reproduction numbers (R_0), reduced generation times, and enhanced binding to host receptors, among other features. Many but not all of the mutations involve spike proteins.[16] One study published in Zenodo, an open repository operated by the European Organization for Nuclear Research (CERN), determined that the novel mutations in the spike protein genes of SARS-CoV-2 variants such as Omicron do not appear to follow a Darwinian trajectory; the author speculated that they were likely designed molecules.[17]

Coronavirus Genomes

The *Coronaviridae* family is divided into four genera based on differences in the viral genomes: alpha, beta, gamma, and delta.[18] Alpha and beta coronaviruses infect mammals while gamma and delta coronaviruses infect primarily birds. The four genera likely emerged from a common ancestor around 300 million years ago, which coincides with the division of amniotes into mammalian and avian species.[19,20] Coronaviruses are large, "pleomorphic" (i.e., various shapes or sizes), spherical particles with symmetrical protein nucleocapsids and lipid bilayer envelopes embedded with membrane proteins, including spike proteins, that interact with host cells.[21]

While linear and non-segmented, coronavirus genomes can be best understood by being divided into two sections. The first section at one end (the 5′ end), constituting about two-thirds of the entire genome, consists of two large open reading frames (ORFs) that encode for fifteen to sixteen nonstructural proteins.[22] Most of these nonstructural proteins are enzymes involved with viral replication, including RNA modification and processing as well as RNA proofreading. The second section of the genome at the opposite end (the 3′ end) consists of ORFs that encode for structural and accessory proteins. The structural proteins constitute the parts of the virus particle, including the nucleocapsid as well as the membrane, envelope, and spike proteins. The accessory proteins are believed to be used for modulating host immune responses to infection and contribute to viral pathogenicity.[23]

Virus-Host Interfaces

Not surprisingly, the most variable section of the coronavirus genome and most prone to mutations is the ORF that encodes for the spike protein, particularly its receptor-binding domain (RBD).[24] The RBD is the part of the spike protein that binds to the receptor of a host cell, like a key fitting into a lock. The better the fit between the key and lock, the better the virus enters the cell like an intruder.

The spike proteins of enveloped viruses, like influenza and coronaviruses, must be activated to fuse the viral and host cell membranes together.[25] Spike proteins are considered "suicide" proteins because they act only once, undergoing irreversible structural changes for the fusion process.[26] Influenza viruses have hemagglutinin (HA) proteins that serve as spike proteins and usually depend on a low pH to induce structural changes.[27,28]

Coronavirus spike proteins use different strategies to be activated. They are composed of two primary subunits called S1 and S2. To be activated, the two subunits are typically split apart at cleavage sites, sometimes at two different locations designated S1/S2 and S2′ (called S2 Prime). Host cell enzymes called proteases perform the cleaving. Enzymes are proteins that facilitate biochemical processes, and one or more host cell enzymes located on the host cell surface or inside the host cell can be involved in the cleavage process.

"Cleave" is a strange verb because it can have opposite meanings. For example, enzymes can cleave two molecules either apart or together. In the case of viral spike protein cleavage sites, "to cleave" generally means that they will be split apart upon entry into cells. But during biosynthesis, in preparation to exit the cells, the cleavage sites are not split apart, but instead are sometimes "primed" for subsequent activation.[29]

SARS-CoV-1

The receptor-binding domain of SARS-CoV-1 binds to a specific host cell receptor called the angiotensin-converting enzyme 2 (ACE2) receptor located on the surface of the cell. The ACE2 receptor has important physiological functions because it is part of the body's renin-angiotensin-aldosterone system (RAAS) that regulates vascular function, such as blood flow and blood pressure. Many organs, including the brain, heart, kidneys, lungs, liver, and intestines, possess ACE2 receptors and are vulnerable to viral infection.[30,31] A mutation of a civet SARS-CoV spike protein involving only a few amino acids in the RBD allowed it to adapt to human airway epithelial cells, increasing its affinity for human ACE2 receptors. This evolution of the RBD facilitated the spillover event from civets to humans.[32]

After binding to ACE2 receptors, the SARS-CoV-1 virus entered host cells in endosomes, temporary membrane-bound vesicles used for cellular transportation. Once inside the cell, the spike protein was cleaved into S1 and S2 subunits by cathepsin L, an endosomal protease. Viral fusion with an endosomal membrane allowed the virus to release its genome into the cell cytoplasm.[33]

The virus could also enter cells without endosomes and instead used proteases at the host cell surface, a process considered to be a hundred- to a thousandfold more efficient for viral entry than the endosomal method.[34] This process was facilitated by the cleavage sites of spike proteins. Importantly, increases in the number of *basic* amino acids in spike protein cleavage sites increased their cleavability. Basic amino acids, such as lysine and arginine, have a neutral pH as opposed to other amino acids, such as aspartic acid and glutamic acid, that have an acidic pH.

Furin is a ubiquitous protease expressed by many cells in all invertebrates and vertebrates. It catalyzes large numbers of precursor protein substrates into their active forms. First identified in 1990 after a twenty-five-year search for a key mammalian enzyme, it was first believed to be an "unglamorous housekeeping protein," but it was subsequently recognized to be involved in many critical cellular processes ranging from embryogenesis to pheromone production as well as in deleterious disease processes, including cancers and dementia.[35,36,37] To activate proteins, furin's preferred cleavage site consists of the amino acid sequence R-X-K/R-R. "R" represents arginine, and "X" can represent any amino acid. "K" represents lysine. The "/" identifies the cleavage site. Therefore, the typical "furin cleavage site" has the amino acid sequence arginine-X-lysine/arginine-arginine.

Many viruses take advantage of the critical role of furin and use it to activate their spike proteins at their cleavage sites.[38,39] Naturally occurring furin cleavage sites with the amino acid sequence R-X-K/R-R have been observed in influenza and coronaviruses.[40,41,42] In the case of highly pathogenic avian influenza, the presence of a furin cleavage site in its hemagglutinin "spike" protein allows it to target many tissues in chickens, and it is considered a major factor for its pathogenicity, particularly when compared to low pathogenic avian influenza, which uses the enzyme trypsin for spike protein activation.[43] SARS-CoV-1 had a cleavage site, but it was cleaved by cathepsin L, not furin.

Understanding the role of cellular proteases like furin and their interactions with coronavirus spike proteins has been an intense area of study since as early as 2005, when researchers in Canada and the United States noted that unlike other coronaviruses, SARS-CoV-1 lacked a furin cleavage site. They created a synthetic furin cleavage site in its spike protein and studied its effects on infected cells with and without enzyme inhibitors.[44] In 2006, researchers at the University of Montana introduced a furin cleavage site into the S1/S2 junction of the SARS spike protein, which potentiated viral membrane fusion activity.[45] Two years later, a team in Japan inserted furin cleavage sites into SARS-CoV-1 spike proteins, also resulting in facilitated viral fusion activity.[46] In 2009, researchers at Cornell University examined SARS-CoV-1 spike proteins along with the spike proteins of other coronaviruses, such as the avian infectious bronchitis virus, which possessed two furin cleavage sites at the S1/S2 junction and at S2′. Since SARS-CoV-1 did not have a furin cleavage site, they inserted two of them into the virus, one at the S1/S2 junction and the other at the S2′ location. These two insertions increased the ability of SARS-CoV-1 to fuse with host cell membranes and enhanced its infectivity.[47] In general, the possession of furin cleavage sites helps make coronaviruses more infectious.[48]

MERS-CoV

In contrast to SARS-CoV-1, MERS-CoV used a different target host cell receptor called dipeptidyl peptidase 4 (DPP4).[49] This receptor is expressed by immune cells, intestinal and renal brush border membranes, vascular endothelium, and glandular epithelial cells, as well as cells in the liver, pancreas, and lungs.[50,51] It plays a key role in activating T cells, part of the immune system, and in regulating the biologic activity of incretin, a hormone involved in the maintenance of normal glucose homeostasis. The DPP4 receptor is also involved in pathological processes including obesity and diabetes, inflammation, tumor biology, immune-mediated diseases, and of course, viral entry.[52]

MERS-CoV combines elements of both SARS-CoV-1 and avian infectious bronchitis virus. Like SARS-CoV-1, it can enter cells via the endosomal route and uses endosomal proteases like cathepsin L for activation. And like the avian infectious bronchitis virus, it has two furin cleavage sites. One is located at the S1/S2 junction and the other at the S2′ location. Both locations have the amino acid sequence arginine-X-X-arginine (R-X-X-R).

Cells with DPP4 receptors and high furin levels have been found to be highly susceptible to MERS-CoV infection. Furin appears to cleave (i.e., split apart) the S2′ site during viral host cell entry, but it also partially cleaves (i.e., "primes") the S1/S2 site during *biosynthesis* of the MERS-CoV spike protein. "Priming" the S1/S2 furin cleavage site means that it is "activated" and ready for future use, namely for cell entry. Researchers speculate that the unusual two-step furin activation process used by MERS-CoV, allowing it to infect many different cell types, might have contributed to its emergence into human populations.[53]

SARS-CoV-2

Like the original SARS-CoV-1, SARS-CoV-2 uses ACE2 receptors for host cell entry, but it binds ten- to twenty-fold more strongly to the receptors compared to SARS-CoV-1. This enhanced binding affinity might contribute to its high infectivity rates.[54,55] SARS-CoV-2 uses two different proteases to activate its spike protein: furin at the S1/S2 site and transmembrane serine protease 2 (TMPRSS2) at the S2′ site.[56] Once separated, the two subunits undergo structural changes to facilitate entry into the host cell.[57] The S1 subunit, which includes the receptor-binding domain, varies in amino acid sequence and length. The S2 subunit is relatively homologous and includes a fusion protein that binds the virus and host cell membranes together.[58] After it is cleaved from the S1 subunit, the S2 subunit anchors its fusion protein into the host cell's target membrane. This process creates a fusion pore that facilitates the viral genome to enter the cell's cytoplasm.[59]

The furin cleavage site of the SARS-CoV-2 spike protein has been a source of controversy because it is not present in other SARS-like coronaviruses called *Sarbecoviruses*. For example, bat coronavirus RaTG13 and pangolin coronaviruses that are closely related to SARS-CoV-2 do not have furin cleavage sites. In addition, the amino acid sequence of the SARS-CoV-2 furin cleavage site is highly unusual, proline-arginine-arginine-alanine-arginine (P-R-R-A-R), and it contributes to the infectiousness of the virus.[60]

Some researchers have argued that the virus was bioengineered with the insertion of a furin cleavage site not previously seen in SARS-like coronaviruses and a receptor-binding domain specifically designed to infect humans through gain-of-function research.[61,62] Gain-of-function research has been defined as research that provides new capabilities to microbes through bioengineering and other methodologies.[63,64] The next chapter will further discuss gain-of-function research, an area of intense debate.

Dimension Three

	Political, Social, and Economic Factors
CHAPTER FIVE	Gain-of-Function Research, Biosafety, Biosecurity, and Bioethics

Gain-of-Function Research, Biosafety, Biosecurity, and Bioethics

Gain-of-Function Research

While biomedical research has resulted in great breakthroughs in medicine and public health, it can present risks. In 2004, the National Research Council of the National Academies of Sciences, Engineering, and Medicine published a seminal report, *Biotechnology Research in an Age of Terrorism*, in response to concerns raised by potentially dangerous experiments, such as a bioengineered mousepox virus that was intended to render mice infertile to control their populations but instead killed them even though they had been vaccinated against the mousepox virus. Another experiment created an infectious polio virus in a laboratory using chemically synthesized oligonucleotides.[1,2,3]

To reduce the risks posed by biomedical research, the report issued a list of recommendations, including seven experiments that should not be done.[4]

The seven experiments would:

1. Demonstrate how to make a vaccine ineffective.
2. Confer resistance to antibiotics or antiviral agents.
3. Enhance a pathogen's virulence or make a non-virulent microbe virulent.
4. Increase the transmissibility of a pathogen.
5. Alter the host range of a pathogen.
6. Enable a pathogen's ability to evade diagnostic or detection modalities.
7. Enable the weaponization of a biological agent or toxin.

In response, the National Institutes of Health (NIH) adopted some of the report's recommendations, such as the creation of a National Science Advisory Board for Biosecurity, but it did not implement others.

Most notably, the funding of potentially dangerous research continued.

In September 2011, Ron Fouchier, the virologist from the Erasmus Medical Center in the Netherlands who had studied CoV-NL63 seven years earlier, presented his research findings on work involving the highly pathogenic avian H5N1 influenza virus at a European Scientific Working Group on Influenza conference in Malta. The studies induced mutations into the virus by forcing airborne transmission between ferrets. Up until that point, the virus had not acquired the capability of airborne transmission between mammals. The researchers wanted to see if it *could* acquire the capability. It was potentially dangerous gain-of-function research, partially funded by the National Institute of Allergy and Infectious Diseases (NIAID) of the NIH. While none of the ferrets died, the research raised alarms throughout the international public health community.[5,6]

Three months later, the global condemnation of the ferret research prompted Drs. Anthony Fauci, Gary Nabel, and Francis Collins, leaders of the NIAID and the NIH, to write an opinion piece in the *Washington Post* defending gain-of-function research. They argued that important scientific and public health questions regarding the likelihood of dangerous mutations arising naturally in viruses, such as highly pathogenic avian H5N1 influenza, remained unanswered.

"Given these uncertainties, important information and insights can come from generating a potentially dangerous virus in the laboratory," they wrote.

The ferret transmission studies were meant to fill knowledge gaps. Although they acknowledged the uncertainties surrounding the risk-benefit ratios of such research, they assured the public that the work was being safely conducted in high-security laboratories.[7]

In 2013, the White House Office of Science and Technology Policy (OSTP) issued a framework for guiding health and human services (HHS) funding decisions on research proposals involving highly pathogenic influenza H5N1 viruses transmissible by respiratory droplets among mammals. Since biomedical research can be used for both benevolent and malevolent purposes, a concept known as "dual use research of concern," OSTP recommended, among other things, that proposed dual use research of concern could be funded if it met certain criteria, including:

The virus generated in the laboratory could evolve naturally.
The research addressed scientific questions with high public health significance.
No feasible less-risky alternative research existed.
Biosafety and biosecurity risks could be sufficiently managed and mitigated.
The research would be supported by funding mechanisms that facilitated appropriate oversight, conduct, and communication.[8]

Biosafety and biosecurity have received much attention over the decades. The next two sections will briefly discuss these issues, focusing on high-containment laboratories and laboratory-acquired infections. Bioethics has traditionally focused on human subjects rather than on biotechnology research. The final section will discuss bioethics as it pertains to biotechnology research.

Biosafety

In the 1970s, genetic bioengineering was a hot new technology in which pieces of DNA from one organism could be inserted into the DNA of another organism. But concerns arose about the potential risks posed by the hybrid organisms escaping from laboratories and causing inadvertent harm to public health and to ecosystems. Paul Berg, a Stanford biochemist and Nobel laureate, played a key role in outlining these concerns, largely because his own work raised alarms because it involved using simian virus (SV40), which can cause tumors in rodents, to introduce new genes into mammalian cells. Ultimately, his goal was to develop new methodologies for gene therapy. As a test run, he inserted segments of DNA from *Escherichia coli* bacteria into the SV40 genome.[9] There were concerns that the novel microbe might escape from the lab and become a cancer-causing agent.[10] Religious leaders representing eighty different faiths and denominations called for a halt to such experiments. Berg decided to postpone his work.[11]

The potential benefits of genetic bioengineering to develop powerful new therapeutics in medicine and new disease- and drought-resistant crops in agriculture were tremendous. Berg and other distinguished scientists convened several meetings at the Asilomar Conference Center in Pacific Grove, California, to address the public's concerns, and they voluntarily agreed to a moratorium on their research until they could figure out how to reduce the risks.[12]

In February 1975, 140 participants, including scientists, journalists, lawyers, and government officials, attended the International Congress on Recombinant DNA Molecules at Asilomar. Heated debate ensued. Berg recalled that a breakthrough occurred when a recommendation was made that different levels of laboratory biosafety could be applied to experiments posing different levels of risk. In other words: the riskier the experiment, the more stringent the laboratory biosafety facilities required.[13]

Experiments posing no risk could be done on an open bench. The riskiest experiments could be done in high-containment laboratories with space suits and air locks. Experiments falling in between these two extremes could be conducted in varying levels of biosafety precautions. The goal was to protect the laboratory workers, the public, and the environment from potential harm posed by the experiments. On July 7, 1976, the US government issued a notice of these recommendations along with public comments in the *Federal Register*.[14,15]

Efforts to improve laboratory biosafety began long before the Asilomar meeting. The American Biological Safety Association (ABSA) held its first unofficial meeting in April 1955 at Camp Detrick in Frederick, Maryland, where fourteen representatives from Camp Detrick, Pine Bluff Arsenal in Arkansas, and Dugway Proving Ground in Utah met to share experiences and knowledge regarding biosafety, laboratory-acquired infections, and other safety-related issues relevant to the US offensive biowarfare program, which had been ongoing since World War II.[16,17] After a series of well-publicized mishaps during open-air tests that affected animals and humans, President Richard Nixon ended the program in 1969 in response to public outrage.[18,19]

Despite the cessation of the offensive biowarfare program, interest in laboratory biosafety grew. Over the years, the ABSA's laboratory biosafety improvement effort was joined by the NIH, the Centers for Disease Control and Prevention (CDC), and eventually by hospitals, universities, private laboratories, and industry. Biological Safety Conferences were held yearly, and the topics of discussion included laboratory-acquired illnesses, biosafety cabinets, and decontamination.

At the 21st Biological Safety Conference held in November 1978 in Scottsdale, Arizona, Ralph Kuehne, a safety officer, described the new special containment facilities at the US Army Medical Research Institute of Infectious Diseases to study highly infectious virulent viruses, such as Ebola, Lassa fever, and Marburg, for which no vaccines or medications were available.[20] Six years later, at the 27th Biological Safety Conference held in October 1984 in Raleigh, North Carolina, Dr. Emmett Barkley, the founder of the NIH Division of Safety in 1979, presented the four biosafety levels (BSLs) that were established in response to the proposed risk assessment protocol discussed at the 1975 Asilomar conference.

The first *Biosafety in Microbiological and Biomedical Laboratories* (BMBL) manual was published in 1984 to serve as the principal reference for biosafety practices in laboratories. The following year, the American Biological Safety Association became official. As of 2022, there have been six editions of the BMBL, the goal of which is to protect laboratory workers, the public, and the environment from exposure to infectious pathogens that are handled and stored in laboratories.[21,22]

Laboratory-Acquired Infections

In general, incidence rates of laboratory-acquired infections (LAIs) are challenging to estimate because they rely on published case reports, which depend on laboratory workers seeking medical care for illnesses that are recognized as possible LAIs. Laboratory confirmation is needed to prove that the illness is an LAI, and the incident must be published in the medical literature. Over a thirty-six-year period, between 1979 and 2015, LAI reports involving only viruses totaled 765 symptomatic and 439 asymptomatic infections. Of the symptomatic infections, approximately 65 percent occurred in research laboratories. Of these, there were 19 deaths, 7 secondary infections, and 5 tertiary infections.[23]

Furthermore, of the total viral LAI reports, 219 were zoonotic and associated with laboratory animal activities, which resulted in 180 seroconversions and 2 fatalities, involving one graduate student and one veterinary student. Most of the zoonotic LAI reports were due to arboviruses (192 symptomatic and 122 asymptomatic) and hantaviruses (189 symptomatic and 74 asymptomatic). Eighty-six percent of the hantavirus LAI reports involved researchers who thought that they were studying uninfected rodents. There were 6 reported SARS-CoV-1 LAIs involving four researchers in China and one researcher each in Singapore and Taiwan. In China, SARS-CoV-1 infectious material was incompletely deactivated when transferred from BSL-3 to BSL-2 facilities, resulting in aerosol exposures. In Singapore, cross-contamination of a West Nile virus with a SARS-CoV-1 virus resulted in an infection in a BSL-2 laboratory. In Taiwan, a researcher wore inadequate personal protective equipment and did not use a disinfectant when cleaning up a leaking biohazard bag in a BSL-4 lab. In many cases, however, the exact cause of an LAI is unknown, but laboratory and wild animals cause many viral LAIs from scratches and bites.[24]

Biosecurity

Biosecurity is defined as the protection of dangerous "biological select agents and toxins" (also known as "select agents" or BSATs) from loss, misuse, or theft.[25] Therefore, a "biosecure" laboratory provides physical security to prevent unauthorized access to these agents, and only conscientious, honest, and reliable individuals can work with them. Historical events such as the 1984 salad bar poisoning with *Salmonella* by members of the Rajneeshee commune in The Dalles, Oregon, to influence a local election and the anthrax letter attacks of 2001 that killed five and sickened seventeen people led to the passage of the USA PATRIOT Act of 2001 and the Public Health Security and Bioterrorism Preparedness and Response Act of 2002, which required the US Departments of Health and Human Services and Agriculture to publish regulations for the possession, use, and transfer of select agents (Select Agent Regulations, 7CFR Part 331, 9 CFR Part 121, and 42 CFR Part 73). These regulations went into effect on February 7, 2003.[26]

The US Federal Select Agent Program oversees the possession, use, and transfer of BSATs that have the potential to cause serious harm to human, animal, or plant health and safety.[27,28,29] Notable pathogens in the BSAT list include "SARS-CoV-1 as well as SARS-CoV-1/SARS-CoV-2 chimeric viruses resulting from any deliberate manipulation of SARS-CoV-2 to incorporate nucleic acids coding for virulence factors."[30]

Since 2015, the CDC and Department of Agriculture (USDA) have been jointly issuing annual reports of the Federal Select Agent Program. The program monitors and reports the numbers of microbial thefts, losses, or releases (a.k.a. exposures) resulting in illness among laboratory workers, the public, or transmission into the surrounding community or environment each year. From 2015 to 2020, there were zero reports of theft, 63 reports of losses, and 1,183 reports of releases, with an average of almost 200 releases per year. Examples of causes of releases include bites or scratches from an infected animal, failure (or problem) with personal protective equipment, needlestick or other percutaneous exposure with a contaminated sharp object, or spill of a BSAT.[31]

In 2020, a total of 158 releases were reported. One release does not necessarily equal one individual exposure. There might be multiple individuals involved in a single reported release. For example, according to the 2020 report, 122 releases involved a total of 483 individuals requiring occupational health services such as medical assessments, diagnostic testing, and/or pharmaceutical prophylaxis. One worker became ill after exposure to *Coxiella burnetii*, the bacteria that causes Q fever, from contact with infected animals, but fortunately fully recovered.[32]

As of 2020, a total of 244 entities were registered with the Federal Select Agent Program. Thirty-five percent of these entities were academic institutions; 29 percent were nonfederal government; 16 percent were commercial; 15 percent were federal government, and 6 percent were private entities. Of the 244 entities, 29 percent possessed BSL-2 labs, 78 percent BSL-3 labs, and 3 percent BSL-4 labs. The geographic locations of the entities were not provided.[33]

The CDC and USDA work with the Federal Bureau of Investigation to identify individuals who apply for access to BSATs and who might pose a security risk. At the end of 2020, a total of 8,121 individuals were approved to have access to BSATs, some of whom worked at several different entities. Twelve individuals who applied for access were identified as "restricted persons" and were denied. Nine of them had been convicted of a crime punishable by imprisonment for a term exceeding one year. One was an alien illegally or unlawfully living within the United States, and one unlawfully used a controlled substance.[34]

Because hundreds of safety lapses occur in US laboratories conducting research on select agents, the US Government Accountability Office (GAO) issued a report in October 2017 to investigate the Federal Select Agent Program and found that it did not fully meet key elements of effective oversight. According to the report, the deficiencies included a lack of independence from the laboratories it oversees, a lack of performance reviews, a lack of transparency, a lack of enforcement, and a lack of technical expertise.[35]

The GAO report compared the United States' program to those in other countries that also have high containment (i.e., BSL-3 and 4) laboratories.[36] For example, Great Britain's Health and Safety Executive is an independent central government agency with the mission to protect worker and public health and safety as well as to oversee laboratories working with pathogens. It avoids the US program's conflict of interest because none of the laboratories that it oversees are part of the same agency.[37]

At the international level, as of 2021, eighty-six states possess or are building high-containment laboratories: fifty-nine are BSL-4 and three thousand are BSL-3.[38] BSL-4 labs are spread over twenty-three countries, with the largest number (twenty-five labs) concentrated in European countries, including Belarus, France, Germany, Italy, the Russian Federation, Sweden, Switzerland, and the United Kingdom. Asia possesses thirteen BSL-4 labs in China, India, Japan, South Korea, Singapore, and Taiwan; North America has fourteen labs in Canada and the United States. Australia has four labs, and Africa has three labs in Côte d'Ivoire, Gabon, and South Africa.[39]

More than three-fourths (46/59) of the BSL-4 labs are in urban centers, which heightens the health security risk in the event of an accidental laboratory-acquired infection or leak. Sixty percent (36/59) are government-run institutions. Most (48/59) focus on human health; seven address animal health, and four study both human and animal pathogens.[40]

Of course, high containment labs are only effective if they are used. Evidence suggests that the SARS-CoV-1 research done in Wuhan, China, was performed in a low-containment BSL-2 lab rather than a high-containment lab.[41,42]

Safety issues at the lab had been a concern.[43] Two years before the pandemic, US embassy officials visited the facility and found biosafety lapses while risky research was being conducted on SARS-like bat coronaviruses capable of infecting human ACE2 receptors. One cable warned that the lab's work posed a risk of a new SARS-like pandemic because of a lack of appropriately trained technicians and researchers.[44]

Each nation is responsible for biosafety and biosecurity within its borders. The United States has issued many biosafety and biosecurity related policies since 2010, including the National Biodefense Strategy in 2018, to counter biological threats to humans, animals, agriculture, and the environment whether naturally occurring, accidental, or deliberate.[45] China began developing its biotechnology industry in the 1980s, but regulations regarding laboratory biosafety and biosecurity, while deliberated in 2018 and 2019, were not passed until after the emergence of the COVID-19 pandemic. China's National People's Congress (NPC) passed the Biosafety/Biosecurity Legislation on October 17, 2020, and the law took effect on April 15, 2021. The legislation uses the term "shengwu anquan," which broadly means "biosafety and biosecurity," and seeks to protect humans, animals, plants, and the environment from inadvertent risks and hazards caused by working with pathogens and toxins as well as by preventing deliberate misuse, theft, or diversion of biotechnology.[46]

There is no single international entity with the mandate to oversee global biosafety and biosecurity. While the mission of the Biological Weapons Convention (BWC) is to prevent the development, acquisition, transfer, stockpiling, and use of biological agents and toxins as weapons, it has no mechanism to enforce the treaty. Its Implementation Support Unit was established in 2006 within the Geneva Branch of the United Nations Office for Disarmament Affairs to provide administrative support to the BWC.[47] (Please see note 19 for further description of the BWC).

The mission of the World Health Organization (WHO) is to promote health, keep the world safe from health emergencies, and serve the vulnerable.[48] The International Health Regulations (IHR), originally adopted by the World Health Assembly in 1969 largely to curb the spread of cholera, plague, and yellow fever, were revised in May 2005 and entered into force in June 2007, after SARS and HIV/AIDS demonstrated the regulations' inadequacies.[49,50] Although the purpose and scope of the revised IHR are "to prevent, protect against, control and provide a public health response to the international spread of disease in ways that are commensurate with and restricted to public health risks, and which avoid unnecessary interference with international traffic and trade," they do not specifically address laboratory biosafety and select agent biosecurity.[51]

In 2004, the United Nations Security Council passed Resolution 1540 to deter states from assisting non-state actors in using, among other weapons, biological weapons for terrorist purposes. The UN Office for Disarmament Affairs supports the Security Council's activities, including facilitating cooperation between subregional, regional, and international organizations, but it is not explicitly involved in the oversight of biosafety or biosecurity of laboratories.[52]

To support the BWC, the revised WHO IHR, the UN Security Council Resolution 1540, and other international health efforts, the Global Health Security Agenda (GHSA) was launched in February 2014 with the objective to enhance each nation's capabilities to prevent, detect, and respond to infectious diseases.[53] GHSA members made commitments to identify and address gaps, build political will and coordination, and invest in health security, among other efforts.[54] However, laboratory biosafety and biosecurity were not specifically mentioned.

Bioethics

Ethics is defined as "a set of moral principles," and bioethics is defined as "a discipline dealing with the ethical implications of biological research and applications especially in medicine."[55] Societies expect biomedical research to be ethical and benevolent. Some gain-of-function studies that augment the capabilities of human and animal pathogens are arguably questionable and raise bioethical concerns.

Historically, the field of bioethics focused on research involving human subjects because of past atrocities committed in the pursuit of medical knowledge. For example, Nazi physicians conducted crimes against humanity in the name of science during World War II. The Nuremberg trials brought these "experiments" to light and led to the Nuremberg Code, which established ten directives for experiments involving human subjects, including that the risks should never exceed the benefits and that the subjects involved should have the legal capacity to give consent.[56,57]

In 1964, the World Medical Association passed the Declaration of Helsinki: Recommendations Guiding Medical Doctors in Biomedical Research Involving Human Subjects, promoting ethical principles for medical research and patient care. One of the principles stated that research protocols must be submitted to independent ethics committees prior to the initiation of a study to decide if the proposed studies meet international standards and norms.[58]

In the United States, public outrage over the infamous Tuskegee Experiment, conducted from 1932 to 1972 by government physicians who withheld lifesaving antibiotics from poor Black men infected with syphilis, prompted Congress to pass the National Research Act of 1974.[59] The act established the National Commission for the Protection of Human Subjects of Biomedical and Behavioral Research. This commission published *The Belmont Report*, which identified basic principles and guidelines to address ethical issues arising from human subject research.[60]

The Code of Federal Regulations (Title 45 Code of Federal Regulations Part 46) provides current ethics policies for federally funded human subject research.[61] Within the Department of Health and Human Services, the Office for Human Research Protections oversees Institutional Review Boards (IRBs), which are independent entities responsible for the oversight of federally funded human subject research, ensuring that it meets the ethical principles established in *The Belmont Report*.[62] For animal research, Institutional Animal Care and Use Committees serve similar purposes.[63]

While biomedical research does not have the lengthy ethics abuse history of human subject research, it has generated concerns. In 2014, in response to the bioethics concerns prompted by the controversial influenza studies in ferrets, the Obama administration imposed a funding pause of MERS, influenza, and SARS gain-of-function research.[64] During this moratorium, the NIH requested that the National Science Advisory Board for Biosecurity (NSABB) and the National Academy of Sciences to help it decide how best to proceed with these types of research studies.[65]

In 2016, the NSABB issued its findings and found, among other things, that only a small fraction of studies, called gain-of-function research of concern (GOFROC), posed very high levels of risk that warranted additional federal and institutional oversight. The scientific merit of these GOFROC studies would need to be evaluated along with other considerations, including ethical, legal, public health, and societal values. For some GOFROC studies, such as those that make a pathogen highly transmissible and/or virulent and likely capable of wide and uncontrollable spread and/or cause significant morbidity and/or mortality in human populations, the NSABB concluded that they "entail significant potential risks and should receive an additional multidisciplinary review prior to determining whether they are acceptable for funding. If funded, such projects should be subject to ongoing oversight at the federal and institutional levels." The benefits, they argued, were new scientific knowledge, countermeasure development, and biosurveillance.[66]

One year later, the White House Office of Science and Technology Policy issued policy guidance for HHS reviews of "potential pandemic pathogen care and oversight (P3CO)." The policy recommendations included, among other things, that an independent expert review process should determine if the proposed research was scientifically sound, ethically justifiable, and whether the benefits to society outweighed the risks of a pandemic.

The recommended policy guidelines further stated: "The pathogen that is anticipated to be generated by the project must be reasonably judged to be a credible source of a potential future human pandemic. An assessment of the overall potential risks and benefits associated with the project determines that the potential risks as compared to the potential benefits to society are justified."[67] In other words, it's okay to take these risks.

These policy guidelines should raise concerns. *No project* should generate a pathogen that is a credible source of a potential future pandemic. Such a project would meet the National Research Council's criteria of the seven experiments that should not be done.

On December 19, 2017, the NIH announced the end of its moratorium on gain-of-function research involving MERS, influenza, and SARS viruses. On the same day, it released a framework for guiding funding decisions about proposed research involving enhanced potential pandemic pathogens (ePPPs), stating that such research was essential to protecting global health and security. The framework defined an ePPP as resulting from laboratory enhancement of transmissibility and/or virulence of a pathogen. In other words, ePPPs were not naturally occurring.[68]

As of February 28, 2022, according to the NIH, only three projects involving ePPPs were reviewed by the HHS P3CO Review Group after the moratorium ended. Two of the projects, involving the influenza virus, were funded and subsequently completed. A third project involving an unspecified ePPP was ultimately not funded.[69]

Information about NIH's 83,017 grants, totaling over $54.3 billion, awarded across twenty-seven institutes and centers, is publicly available on its RePORTER website. While the NIAID has the second-highest number (9,339) of actively funded grants, after the National Cancer Institute (NCI), which has 10,744 grants, its total grant funding, over $7.8 billion, is larger than the NCI's $7.7 billion.[70]

SARS-CoV-2 Origin Controversy

No intermediate animal host has been identified as the source of SARS-CoV-2, and no serological studies of humans have been published confirming a natural spillover event. Circumstantial evidence suggests that the virus resulted from gain-of-function research and escaped from a laboratory, possibly as a laboratory-acquired infection. The Wuhan Institute of Virology (WIV) is a leading global center for coronavirus research.

Dr. Peter Daszak, president of EcoHealth Alliance in New York City, served as the project leader of several NIAID grants with funding from June 2014 to June 2022 to support coronavirus research at the WIV. The research included collecting bat samples, screening coronavirus genomes, sequencing spike proteins to identify coronaviruses with high spillover potential, and infecting cell cultures from different species and humanized mice, among other activities. Daszak noted that they found many SARS-like coronaviruses in bats in southern China that used human ACE2 receptors to enter cells. The total cost of the two grants was $1.23 million.[71]

Shi Zheng-li, the lead expert on bat coronaviruses at the WIV, made frequent trips to bat-filled caves in southern China to collect coronaviruses. She worked with Ralph S. Baric, a prominent coronavirus expert at the University of North Carolina, to conduct gain-of-function studies by combining mouse-adapted SARS-CoV-1 with a spike protein from a bat coronavirus named SHC014-CoV to create a chimera, a virus possessing genes from two different viral strains, that could infect human ACE2 receptors.[72,73] The research prompted debate over whether the project's pandemic potential risks outweighed the perceived benefits of predicting future pandemics.[74]

One analysis of the SARS-CoV-2 genome suggested that it was a chimera with much of its genome identical to bat coronavirus RaTG13 and its receptor binding domain almost identical to a pangolin CoV optimized to bind to ACE2 receptors on human cells.[75] These findings led some to conclude that SARS-CoV-2 came from gain-of-function research.[76]

On December 9, 2019, during an interview between Daszak and Dr. Vincent Racaniello, a professor of microbiology and immunology at Columbia University, on *This Week in Virology* on Microbe TV, Daszak stated:

> We did surveillance of bats in southern China, and we've now found . . . over 100 new SARS-related coronaviruses, very close to SARS, some of them get into human cells in the lab, some of them can cause SARS disease in humanized mice models and are untreatable with therapeutic monoclonals and you can't vaccinate against them with a vaccine . . . so these are a clear and present danger. . . . You can manipulate them in the lab pretty easily. . . . Spike protein drives a lot of what happens with a coronavirus zoonotic risk . . . so you can get the sequence, you can build a protein, and we work with Ralph Baric at UNC to do this . . . insert into the backbone of another virus and do some work in the lab.[77]

In April 2020, the NIH canceled an EcoHealth Alliance grant to support coronavirus research at the WIV but reinstated it four months later, along with multiple demands, including that an inspection of the WIV by US federal officials be arranged and that vials of the SARS-CoV-2 sample used to determine its genetic sequence be submitted. Daszak reportedly complained about being caught in political crosshairs and called the demands "heinous." Zheng-li called them "outrageous."[78]

Daszak's denials and protestations, and the NIH's actions, led to a *Washington Post* editorial on October 25, 2021, stating that there were too many unanswered questions regarding the origin of SARS-CoV-2. The editorial noted that Daszak did not disclose a 2018 proposal to the Defense Advanced Research Projects Agency for funding to conduct the bioengineering of spike proteins of chimeric viruses that would make them infect human cells like SARS-CoV-2. In 2019, the WIV removed its online databases of viruses. The editorial board called for Daszak to testify before Congress. He was receiving federal grant funding and must be transparent and held accountable for the research being done with it.[79]

The 2022 election results returned control of the US House of Representatives to the Republicans.[80] On Monday, January 9, 2023, they commissioned a special investigative panel to determine the origin of the virus by questioning scientists as well as former and current federal officials.[81] On the night of November 14, 2023, the Republican-controlled House of Representatives approved a ban of all HHS-funded gain-of-function research that modified risky pathogens, making them potentially more dangerous to people. Biomedical research groups voiced their objections and were hopeful that the Democratic-controlled Senate would not pass similar legislation.[82]

Public distrust of science has been a longstanding issue, and questions about the origins of SARS-CoV-2 have put virology in the crosshairs.[83,84] The biomedical research enterprise requires public trust, and with potentially billions in jeopardy if public trust is lost, the stakes couldn't be higher.

CHAPTER SIX

Conclusions and Policy Recommendations

Coronaviruses have existed for hundreds of millions of years and are part of Earth's microbial ecosystems. We must learn to live sustainably with these microbes to reduce the risks that they pose to humans and animals. To achieve this goal, we must first understand them. Using a multidimensional One Health matrix as a framework, this book has examined coronaviruses in a concise, organized, and systematic fashion. Briefly reviewing the findings of this examination facilitates the development of policy recommendations, which are discussed below.

Animals and Humans

In the early 1930s, veterinarians first identified a novel epizootic in the American Midwest. The epizootic was fast moving and highly fatal to young chicks. It was named the "avian infectious bronchitis virus" and was subsequently identified to be a coronavirus. In the succeeding decades, coronaviruses caused many deadly outbreaks in a wide variety of domesticated animal species, including pigs, turkeys, cats, dogs, horses, cattle, and laboratory mice. These viruses infected many organ systems, including pulmonary, gastrointestinal, and neurological. Veterinarians became experts in coronaviruses because their education, training, and practices included zoonotic pathogens.

Not until the 1960s did medical researchers discover that some upper respiratory tract infections (e.g., common colds) were caused by coronaviruses. They assumed that these viruses were nothing more than mere nuisances since they appeared to only cause colds. The medical community ignored the deadly coronavirus epizootics involving animals. Professional communications and collaborations between veterinarians and physicians were lacking.

Beginning in 2002, the assumption that coronaviruses were not deadly to humans was shattered by the emergence of SARS-Cov-1, followed ten years later by MERS-CoV and almost twenty years later by SARS-CoV-2 (COVID-19). Both SARS-CoV-1 and MERS-CoV had strong evidence to be natural spillover events from two convincing lines of evidence: (1) the viruses and/or antibodies to the viruses were identified in both humans and animals, and (2) humans with occupational exposures to animals exhibited higher rates of seropositivity to the viruses compared to the general human population. In contrast, these lines of convincing evidence were missing with SARS-CoV-2.

The emergence of SARS-CoV-1 and MERS-CoV prompted investigations into understanding the ecology of coronaviruses. The reservoir host species turned out to be certain bats. Bats are an ancient species with long lives, the ability to fly, and unique immune systems. They eat diverse diets, roost in large numbers, hibernate, fill many ecological niches, and most notably, harbor many deadly zoonotic pathogens. In some countries, people eat bats and other wildlife, increasing zoonotic spillover risk.

Much has been written about the dangers of capturing, selling, trading, butchering, and eating wild animals. Housing large numbers of them in live animal markets where they can transmit microbes to each other and to humans poses disease risks. In general, humanity's demand for meat and other animal proteins drives many of the factors that promote naturally occurring zoonotic disease emergence and spread. Domesticated animals are not exempt from this problem since raising hundreds or thousands of them in densely packed quarters also facilitates pathogen transmission.

Consuming less meat and animal proteins, which would lower demand and reduce the production and trade of animals, might reduce the problem. Of interest, India, which has the highest percentage of vegetarians in its population, around 38 to 40 percent, has not had a coronavirus spillover event like China, despite also having live animal markets.[1]

It's unrealistic, however, to impose vegetarianism on the world's populations since global demand for meat and animal proteins has been increasing[2] despite the zoonotic disease risks and other negative externalities, such as greenhouse gas emissions from large-scale, intensive livestock production. A more realistic approach might include policies, procedures, and enforcement mechanisms to protect wildlife and their ecosystems and to change the societal and cultural norms that promote meat consumption.[3]

Altering societal and cultural norms is not easy, however, since behaviors are influenced by many factors, including personal health concerns.[4] Peer pressure influences behavior and might provide an important strategy for policy makers.[5]

Perhaps the emergence of SARS-CoV-1 and MERS-CoV would have been less of a shock if the medical community had been more aware of the history of coronavirus epizootics in animals. Since most pandemic potential pathogens are zoonotic, it behooves the medical and veterinary medical communities to increase communication, collaboration, and cooperation through joint educational, clinical, public health, and research efforts.[6] Medical and veterinary medical schools should teach One Health concepts to their students, emphasizing the fact that we live in a microbial world. It's imperative that health professionals understand the microbial ecosystems their patients live in and that live inside (and on the surface of) their patients.

Public health, agriculture, and wildlife organizations must work together to reduce the risks of zoonotic disease spillover events while ensuring food safety and security. As of January 2023, more than 8 billion humans inhabit Earth.[7] The United Nations estimates that the world population will reach 9.8 billion by 2050 and 11.2 billion by 2100.[8] Figuring out how to sustainably feed ourselves on a warming planet with diminishing natural resources will require interdisciplinary One Health efforts.[9]

Comparing SARS-CoV-1, MERS, and SARS-CoV-2 Human Infections

Similarities and differences existed in the clinical manifestations of SARS-CoV-1, MERS-CoV, and SARS-CoV-2 infections in humans. All three coronaviruses could present with fevers and cough, but asymptomatic cases could also occur. Notably, asymptomatic, communicable cases were most prevalent with SARS-CoV-2, which helped fuel the silent spread of the virus.

All three viruses could infect many organs, but SARS-CoV-2 was unique in that it caused a loss of taste and smell, a variety of rashes and other dermatologic abnormalities, and an extensive list of chronic conditions involving many organ systems. Investigations into the causes and potential treatments of "long COVID-19" continue.

Of the three viruses, SARS-CoV-2 had the lowest case fatality rate, approximately 2 percent, depending on the country, compared to SARS-CoV-1 (9.6 percent) and MERS-CoV (34.5 percent), although unreported asymptomatic cases of SARS-CoV-1 and MERS-CoV might have resulted in artificially high rates. Nevertheless, the next coronavirus pandemic might not be at the lower end of the case fatality rate spectrum.

Environments and Ecosystems

While SARS-CoV-1 and MERS-CoV were capable of airborne spread, this mode of transmission was the exception rather than the rule. In contrast, SARS-CoV-2 spread primarily by the airborne route, which facilitated its communicability, particularly in indoor settings. These differences in communicability were evidenced by the three coronaviruses' different estimated basic reproductive numbers, R_0, especially the SARS-CoV-2 variants, as noted in appendix 2. Superspreading events occurred with all three coronaviruses but were more common with SARS-CoV-2.

Many countries, including the United States, have had a poor track record of preventing respiratory pathogens such as influenza and SARS-CoV-2 from spreading.[10] Surveillance of the quality of indoor air is generally not done, and societies have had a poor track record of reducing indoor respiratory pathogen spread. Arbitrary divisions such as 5 microns for respiratory particle sizes and vague definitions such as droplets, droplet nuclei, and aerosols resulted in baffling and inconsistent public messaging.[11]

There has been scant research on airborne disease transmission, largely because the research is challenging to do and hasn't had the needed support.[12] Collecting air samples for testing is difficult as opposed to water and food sampling. As a result, surveillance of respiratory diseases, such as influenza, have relied on laboratory tests from people exhibiting flu symptoms rather than on indoor air samples. This gap in surveillance resulted in a data vacuum during the COVID-19 pandemic. Over a year after the SARS-CoV-2 virus first appeared in Wuhan, China, public health officials continued to argue whether the disease was airborne.

In addition to public health officials, architects and engineers have generally ignored the issue of respiratory pathogen spread when designing and constructing buildings or other indoor environments. Academic and corporate leaders have generally ignored the annual economic costs and health impacts of respiratory diseases such as influenza. Government officials have ignored the threat of respiratory disease spread in buildings, and as a result, building regulations for indoor air quality are typically weak or nonexistent. The impact of the COVID-19 pandemic on all aspects of society should serve as a wake-up call for the importance of preventing indoor respiratory disease transmission.[13]

Governments should ensure that indoor settings reduce airborne disease spread. The use of high efficiency particulate air (HEPA) filters in heating, ventilation, and air-conditioning systems would improve air quality and reduce respiratory disease spread. Modern airliners recycle air and circulate fresh air using HEPA filters, so coronavirus spread can be low, but air quality on subways can be highly polluted.

In contrast to the challenges with indoor air quality monitoring, wastewater monitoring of SARS-CoV-2 became an important surveillance strategy since the coronavirus could be expelled in urine and feces. This monitoring system proved to be extremely useful for gauging viral activity in communities and should be continued for SARS-CoV-2 as well as for other similarly excreted pathogens.

Coronavirus Molecular Biology

Coronaviruses are single-stranded RNA viruses that are notorious for high mutation rates and for the evolution of new variants. The original SARS-CoV-2 strain mutated into many variants of concern that were more communicable and, in some cases, more virulent. Many of the mutations involved the spike proteins that interact with host cell receptors. Both SARS-CoV-1 and SARS-CoV-2 bind to host cell ACE2 receptors that are located on many organs, including the lungs, heart, brain, kidneys, and intestines. But SARS-CoV-2 bound ten- to twenty-fold more strongly to the receptors compared to SARS-CoV-1, making it more capable of infecting cells.

The original SARS-CoV-1 virus and other similar Sarbecoviruses did not have furin cleavage sites in their spike proteins. In contrast, SARS-CoV-2 did have a furin cleavage site in its spike protein, likely contributing to its infectiousness. Some researchers have argued that the unusual furin cleavage site of SARS-CoV-2 was inserted through gain-of-function research, a form of bioengineering.

Gain-of-function research is not new. Since 2005, researchers in the United States and abroad have been inserting furin cleavage sites into SARS-CoV-1 spike proteins to see what would happen. The SARS-CoV-2 furin cleavage site, and the location where the pandemic first appeared, Wuhan, China, site of the Wuhan Institute of Virology, fueled debate as to the possible laboratory origins of the virus.

Biosafety, Biosecurity, and Bioethics

No laboratory is infallible, and evidence shows that laboratory-acquired infections (LAIs) can and do occur, sometimes with lethal consequences. Inherent weaknesses exist in many biosafety oversight regimes, including in the United States. Before the emergence of SARS-CoV-2, biosafety concerns existed at the Wuhan Institute of Virology.

Relying on publications in the medical literature has been an inefficient way to monitor LAIs in biomedical research and clinical laboratories. Personnel who work with active microbial agents should report their exposures if they are aware of them. The list of the agents that they work with should be made available to the relevant state and local medical and public health authorities.[14] This sharing of information would allow health officials to prepare for possible secondary or tertiary spread of laboratory-acquired infections.

All nations are responsible for laboratory biosafety and biosecurity within their borders. Some countries have better oversight capabilities than others. In the United States, the Select Agent Program, a monitoring system jointly run by the Centers for Disease Control and Prevention and the Department of Agriculture for reports of theft, losses, or "releases," another term for pathogen exposures, has issued annual reports since 2015. The program oversees work on select agents in a variety of institutions, including academic, federal government, nonfederal government, commercial, and private laboratories.

A single "release" can involve multiple individuals. On average, there have been almost two hundred releases per year in the United States, many of which required medical attention. Despite the Federal Select Agent Program's laboratory oversight efforts, however, the US Government Accountability Office found many deficiencies in the program, including conflicts of interest, a lack of enforcement capabilities, a lack of transparency, and a iack of technical expertise.

Great Britain's Health and Safety Executive program is an independent central government agency, separate from any of the laboratories it oversees, which enables it to avoid conflicts of interest. Only after the COVID-19 pandemic emerged did the Chinese government pass biosafety and biosecurity oversight legislation.

Eighty-six countries possess high-containment laboratories, many of which are in Europe, Asia, and North America, with a few in Africa and Australia. Over three-fourths are in urban centers, which increases population health security risks in the event of accidental releases.

No single international entity has the mandate to oversee global laboratory biosafety and biosecurity. The Biological Weapons Convention could expand its mission to include laboratory biosafety and biosecurity, but it lacks enforcement capabilities. Neither the World Health Organization nor the UN Office for Disarmament Affairs have been explicitly involved in laboratory biosafety and biosecurity. One solution might involve a collaboration between WHO and the Biological Weapons Convention Implementation Support Unit to create a biosafety and biosecurity laboratory surveillance system based on harmonized data collected and reported at national levels. This collaboration would send the message that the international community takes laboratory biosafety and biosecurity seriously. Another strategy could be to establish an independent, global organization to oversee laboratory biosafety and biosecurity such as the International Biosecurity and Biosafety Initiative for Science that was created in early 2023 by the Nuclear Threat Initiative, a Washington DC–based nonprofit global security organization.[15] Of course, political leaders of all nations would have to provide their support for the success of any collaborative oversight effort.

In general, bioethics has focused on human subject research because of historical abuses committed in the name of science. Bioethics must include biomedical research. The scientific merit of gain-of-function research involving pandemic potential pathogens such as SARS-CoV-1 and MERS-CoV has been debated. While some scientists argue for its importance, others consider it highly unethical and dangerous.

Manipulating viruses to see if they can cause pandemics meets the criteria of an experiment of concern that should not be done, as stipulated by the National Research Council's report, *Biotechnology Research in an Age of Terrorism*.[16] Bioengineering SARS- or MERS-related coronaviruses that would increase their efficient entry into human cells by experimenting on humanized mice could enhance virulence, increase transmissibility, potentially alter host range, and evade available therapies or vaccines. National Institutes of Health documents show that EcoHealth Alliance and its Wuhan Institute of Virology partners conducted such research.[17] The government documents stated that genetically altered mice, expressing human ACE2 or DPP4 receptors, were developed to be experimentally infected with either chimeric SARS-CoV-1 or MERS-CoV, respectively. Some of these chimeric coronaviruses possessed wild type or mutant spike proteins.[18] Mice experimentally infected with some of the SARS-CoV viruses experienced about 20 percent weight loss and had high viral loads in their lung tissues. The experiments demonstrated varying levels of viral pathogenicity.[19]

EcoHealth Alliance and its Wuhan Institute of Virology partners' goal was to be able to predict pandemics, but by doing gain-of-function research, they might have inadvertently created the COVID-19 pandemic. There are less risky ways to predict pandemics than conducting gain-of-function research on pandemic potential pathogens. Predicting the evolution of coronavirus spike proteins could be done using computer models, artificial intelligence, and machine learning rather than genetically manipulating viruses in laboratories.[20,21]

Conclusion

The exact origin of the COVID-19 pandemic remains controversial. So far, there is no serological evidence for a natural spillover event with SARS-CoV-2 like there was for SARS-CoV-1 and MERS-CoV. There is suspicion that gain-of-function research on chimeric SARS-like coronaviruses using humanized mice that expressed ACE2 receptors resulted in a laboratory-acquired infection that set off a chain reaction around the world.

Importantly, not all gain-of-function research is dangerous or problematic. For example, giving bacteria such as *E. coli* and transgenic plants the capability to produce drugs such as insulin is beneficial gain-of-function work. These life forms do not naturally have these capabilities.[22] This work is essential for producing affordable therapeutics; it should be supported and continued.

Public support for science has been predicated on the understanding that scientific research benefits societal well-being. To continue this support, science must be honest, truthful, transparent, and forthcoming with the public. Confusing and misleading information generates mistrust.

Ultimately, a judicious approach to preventing pandemics should cover both natural and unnatural spillover events. Meeting humanity's demand for animal proteins must consider the inherent zoonotic risks and devise strategies to minimize them. Gain-of-function research on pandemic potential pathogens is highly dangerous and unethical and should be banned globally. Enforcement of such a ban must have political support. Humanity doesn't need to help nature create more deadly pathogens. A One Health approach recognizing the interconnectedness between humans, animals, plants, microbial ecosystems, and changing environments is essential for global sustainability and survival. These issues should be of concern to us all.

Acknowledgments

First, I would like to thank Dr. Kenneth McIntosh, professor of pediatrics in the Division of Infectious Diseases at Boston Children's Hospital, who described the history of the early coronavirus research, including his own at the National Institutes of Health during the 1960s. Dr. Katalin Karikó, biochemist and mRNA researcher, graciously responded to my email inquiries with useful resources and comments. Dr. Gary Whittaker, professor of virology at the Cornell University College of Veterinary Medicine, answered my questions about the workings of furin cleavage sites. My Princeton University students from fall 2020 and spring 2022 assisted me with researching the history of coronavirus outbreaks in animals. Finally, Robin Coleman, my Hopkins Press editor, has been a source of support as well.

Most importantly, I could not have written this book without the tireless efforts, comments, edits, and suggestions by my mother, Gladys S. Kahn. Her contributions were invaluable, and I'm eternally grateful to her. She has been my lifeline throughout my life. Thanks to advances in cancer therapies, disease-preventing medications, and vaccines from breakthroughs in biomedical research, she's living an active, independent life.

Appendix 1a

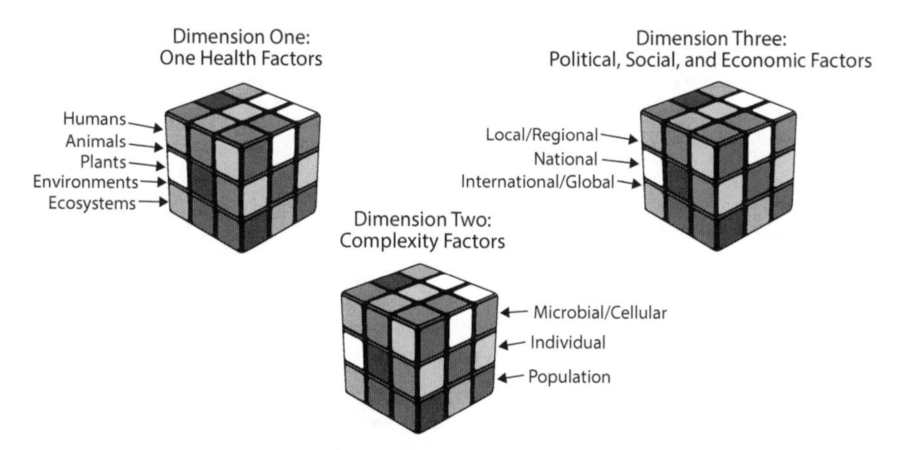

Figure 1. One Health Multidimensional Matrix Tool

Appendix 1b

Brief Timeline of Coronavirus Outbreaks/Discoveries in Animals

Year	Host Animal	Virus/Disease	Transmission	Disease Severity
1931	Baby chicks	Infectious bronchitis virus (IBV) in American Midwest	Not available	40–90% mortality
1946	Swine	Transmissible gastroenteritis virus (TGEV) in American Midwest	Possibly fecal-oral or airborne	80–100% in newborn pigs
1947	Murine (mice)	Mouse hepatitis virus (MHV) in Boston	Experimental	Animals sacrificed in studies
1951	Turkeys	Unfamiliar turkey disease in Washington State	Not available/Difficult to transmit experimentally	15–50% mortality
1957	Swine	Porcine hemagglutinating encephalomyelitis virus (PHEV) in Ontario, Canada	Not available	In very young piglets, approximately 100%
1962	Swine	Porcine Hemagglutinating virus producing encephalomyelitis (PHEV) in baby pigs in Ontario, Canada	Experimental	Very high in young piglets
1963	Feline	Feline infectious peritonitis (FIP) recognized at Angell Memorial Animal Hospital in Boston, determined to be a feline coronavirus (FCoV) several years later	Not available	Not available
1968	Avian, Murine, Human	Coronaviruses recognized as new group	Not applicable	Not applicable
1970	Rat	Rat coronavirus (RCV) discovered in wild and domestic rats. Related to mouse hepatitis virus	Experimental intranasal inoculation of newborn Fischer strain rats	Almost 100% mortality in 1-day-old Fischer rats but 10–25% mortality in 1-day-old Wistar rats

1970s	Turkey	Turkey coronavirus (TCoV) linked with enteritis of turkeys called bluecomb and other intestinal disorders	Fecal-oral spread or dropping-contaminated fomites. Spreads rapidly through flocks. Infected birds shed virus several weeks after recovery from disease. Insects and other animals can serve as mechanical vectors of droppings	Mortality rate depends on host age and comorbidities
1971	Canine	Coronavirus in military dogs with diarrhea (virus related to TGEV) in US army base in Germany	Fecal-oral	Self-limited gastroenteritis with diarrhea
1971	Swine	Porcine epidemic diarrhea virus (PEDV) appears in United Kingdom	Fecal-oral, contaminated fomites, fecal-nasal (airborne via aerosolized particles)	80–100%, especially in piglets
1972	Bovine	Bovine coronavirus (BCoV) infects upper and lower respiratory tracts and intestines. First detected in Nebraska. Also infects horses, buffalo, camelids, deer, elk, and giraffe. Ubiquitous in cattle. Worldwide spread	Viral shed in feces and nasal secretions. Fecal-oral and respiratory (including aerosol) spread. Asymptomatic and recurrent infections are common	Severity of disease varies with host age and immune status. Most calves recover. BCoV winter dysentery causes 1–2% mortality in adult herds
1978	Swine	PEDV found in 4 swine breeding farms in Belgium	Not specified, presumably fecal-oral	Watery diarrhea caused 50% mortality rate in piglets. Decreased rate with increased age
1979	Feline	Feline infectious peritonitis identified as a coronavirus	Fecal and urine spread	Not specified

(*continued*)

Brief Timeline of Coronavirus Outbreaks/Discoveries in Animals *(continued)*

Year	Host Animal	Virus/Disease	Transmission	Disease Severity
1984	Swine	Porcine respiratory coronavirus (PRCV) appears and spreads in Europe. Spreads to United States a few years later	Aerosol and direct contact spread. Viral tropism for respiratory cells. Enteric cells to a lesser extent. Low pathogenicity, fever, and anorexia main signs of disease	Mild or subclinical disease
1988–2001	Avian	29 variants of infectious bronchitis virus (IBV) identified from commercial flocks in California	Respiratory secretions and fecal-oral spread	Vaccines can be effective in controlling disease, but new variants pose challenges
2000	Equine	Coronavirus isolated from diarrheic foal in North Carolina. Coronavirus-like viruses had been previously identified in foals and adult horses. Genetically similar to bovine coronaviruses. Caused dozens of outbreaks in adult horses in United States during the succeeding years	Epidemiology and spread of virus not mentioned	Foal recovered. A foal in another report deteriorated and was euthanized. Approximately 7% of ill horses infected with equine coronavirus die
2000s	Avian	Infectious bronchitis virus has many genotypes. Global spread causing major economic losses	Aerosol transmission. Droppings and nasal discharge also infectious. Virus involves primarily respiratory, renal, and reproductive systems but can affect many tissues	High mortality rate in 1-day-old chicks. Reduced egg production by up to 50%. Long-term infections possible
2005	Canine	Novel, highly pathogenic variant of canine coronavirus in pet shop outbreak in Bari, Italy	Fever, lethargy, vomiting, hemorrhagic diarrhea, and neurologic signs. Virus-induced cytopathic effects seen in all tissues except brain	Highly lethal, especially to young puppies

2010	Swine	Porcine epidemic diarrhea virus (PEDV) in China. Outbreak in more than 10 provinces and more than 1 million piglets die	Fecal-oral transmission. Virus present in sow's milk	High mortality rate in suckling piglets
2013	Swine	Porcine epidemic diarrhea virus (PEDV) in Iowa swine farms. First appeared in United Kingdom in 1971, causing outbreaks around the world. US strain 99% identical to Chinese strains circulating 2011–12	Fecal-oral. Explosive epidemics of diarrhea and vomiting. Over 1 year, spread to most swine-producing areas, causing loss of 7 million pigs	90–95% mortality within 2–3 days in piglets
2014	Swine	Porcine deltacoronavirus emerges in China and Ohio, US	Similar to PEDV	Similar to PEDV
2016	Swine	Swine acute diarrhea syndrome coronavirus (SADS-CoV), also known as porcine enteric alphacoronavirus (PEAV), emerged in China from HKU2-related bat coronavirus	Fecal-oral transmission	90–100% mortality in newborn piglets
2019–2022	Many domestic and wild species, including cats, dogs, tigers, lions, gorillas, snow leopards, spotted hyena, ferrets, mink, mule and white-tailed deer	SARS-CoV-2	Transmission through close contact with humans. Zoonotic transmission from animals to humans appears low. As of May 2, 2022, 362 confirmed cases in companion animals, 18 mink farms affected, 25 states with wildlife cases	Not available

Appendix 2

Environmental Transmission

	SARS-CoV-1	MERS-CoV	SARS-CoV-2 (COVID-19)	Delta variant	Omicron variant
Estimated R_0	1.7 to 3.6	0.45 to 8	2.5 to 6.1	3.2 to 8	5.5 to 24
Primary mode of spread	Respiratory droplets	Respiratory droplets	Airborne (indoors)	Airborne (indoors)	Airborne (indoors)
Airborne transmission	Yes	Yes	Yes	Yes	Yes
Wastewater detection (urine and feces)	Yes	Yes	Yes	Yes	Yes

Appendix 3a

Comparing SARS, MERS, and COVID-19 Clinical and Laboratory Manifestations

Manifestations	SARS-CoV-1	MERS-CoV	SARS-CoV-2 (COVID-19)
Incubation period	2–16 days (median 6 days)	2–14 days (median 5 days)	2.2–11.5 days (median 5 days)
Asymptomatic cases	Yes	Yes	Yes
Initial symptoms	Fever, cough, myalgia	Fever, chills, cough	Fever, cough, fatigue
Respiratory	Dyspnea, respiratory distress, pneumonia, respiratory failure, acute respiratory distress syndrome (ARDS)	Dyspnea, pneumonia, acute respiratory failure, ARDS	Hypoxemia, tracheobronchitis, pneumonia, acute respiratory failure, atypical ARDS
Chest imaging	Radiographs: air-space consolidation, unilateral focal or multifocal or bilateral involvement. CT scans: ground-glass opacification in periphery of affected lung	Radiographs: unilateral or bilateral patchy opacities, consolidation, interstitial infiltrates, pleural effusions	CT scans: bilateral multiple lobular and subsegmental areas consolidation; bilateral ground-glass opacities
Cardiovascular	Tachycardia, hypotension, bradycardia, transient cardiomegaly, acute myocardial infarction, acute coronary syndrome	Acute myocarditis, heart failure, hypotension, shock	Arrythmias, acute myocardial injury, myocarditis, cardiomyopathy
Cardiac enzymes	Elevated creatine kinase	Elevated cardiac enzymes	Elevated troponin I
Gastrointestinal	Nausea, vomiting, watery diarrhea	Vomiting, diarrhea	Loss of appetite, nausea, vomiting, diarrhea, abdominal pain
Serum electrolytes/miscellaneous	Hyponatremia, hypokalemia, elevated lactate dehydrogenase	Elevated lactate dehydrogenase	Hyponatremia, hypochloremia, hypocalcemia
Liver function	Elevated alanine aminotransferase	Elevated alanine aminotransferase, elevated aspartate aminotransferase	Elevated aspartate aminotransferase

Renal function	Acute renal impairment	Acute renal injury, acute renal failure	Acute renal failure
Immune function/inflammation	Leukopenia, lymphopenia, elevated proinflammatory cytokines	Leukopenia, lymphopenia, elevated proinflammatory cytokines	Increased neutrophils, lymphopenia, elevated proinflammatory cytokines. Infected monocytes/macrophages activate inflammasomes, leading to systemic inflammation
Hematologic function	Thrombocytopenia, prolonged activated partial-thromboplastin time, normal prothrombin time, elevated D-dimer levels	Thrombocytopenia	Thrombosis, thrombocytopenia, elevated prothrombin time, elevated D-dimer levels
Nervous system senses	Dizziness, headache, seizures, cerebrovascular disease	Neuropathy, delirium, acute cerebrovascular disease, confusion, seizures	Loss of smell and taste, impaired consciousness, acute cerebrovascular disease, seizures, encephalitis, meningoencephalitis, ischemic stroke, hemorrhagic stroke
Psychiatric	Confusion, depression, anxiety, impaired memory, insomnia	Confusion, depression, anxiety, impaired memory, insomnia	Anxiety, depression, insomnia, post-traumatic stress disorder (PTSD)
Dermatologic	No reports of dermatologic manifestations	No reports of dermatologic manifestations	Erythematous rash, vesicular lesions, urticarial lesions, pseudo-chilblains, acro ischemia and livedoid lesions, maculopapular lesions, among others

(*continued*)

Comparing SARS, MERS, and COVID-19 Clinical and Laboratory Manifestations *(continued)*

Manifestations	SARS-CoV-1	MERS-CoV	SARS-CoV-2 (COVID-19)
Variants	Some	Some	Many
Treatments given	Supportive, steroids, remdesivir	Supportive, remdesivir	Supportive, steroids, remdesivir, convalescent plasma, anti-SARS-CoV-2 monoclonal antibodies, hyperimmune globulin, Paxlovid
Vaccines	No	No	Yes
Chronic sequelae	Psychiatric illness, chronic fatigue syndrome, reduced pulmonary function, hip osteonecrosis from steroid therapy	Chronic fatigue, anxiety, depression, PTSD	General: joint pain, peripheral limb ischemia, neuropathy. Neurologic: brain fog, loss of smell/taste, encephalopathy, stroke. Pulmonary: cough and dyspnea, chest pain. Cardiovascular: palpitations, heart failure, myocarditis/pericarditis. Cutaneous: "COVID toes," urticarial rash, erythematous rash. Psychiatric: PTSD, depression, anxiety
Global case fatality rate	9.6%	34.5%	2% Varies by country, population demographics, time, socioeconomic factors, and national public health efforts, including vaccine availability

Appendix 3b

Alpha and Beta Coronavirus Discoveries in Humans
(Order: Nidovirales, Suborder: Coronavirineae)

Family: Coronaviridae	Year discovered	Genus	Transmission pathway	Human receptors / furin cleavage site
HCoV-B-814	1965	Not classified	Not specified	N/A
HCoV-229E	1966	*Alphacoronavirus*	Bats to camelids/ alpacas, to humans	Amino-peptidase N/No
HCoV-OC 43	1967	*Betacoronavirus*	Rodents to cattle to humans	9-O-acetylsialic acids / Yes
SARS-CoV-1	2002	*Betacoronavirus*	Bats to palm civets to humans	ACE2*, C-type lectin / No
HCoV-NL63	2004	*Alphacoronavirus*	Bats to unknown intermediary to humans	ACE2, heparin sulfate / Yes
HCoV-HKU1	2005	*Betacoronavirus*	Rodents to unknown interme-diary to humans	9-O-acetylsialic acid / Yes
MERS-CoV	2012	*Betacoronavirus*	Bats to dromedary camels to humans	DPP4**, sialic acid / Yes
SARS-CoV-2	2019	*Betacoronavirus*	Bats to unknown intermediary to humans	High affinity to ACE2 receptors / Yes

*Angiotensin-converting enzyme (ACE2)
**Dipeptidyl-peptidase 4 (DPP4)

Appendix 4

Discovering Messenger RNA and the Messenger RNA Vaccines

The discovery of messenger RNA (mRNA) occurred through the efforts of many individuals working collaboratively for over a decade. Two separate teams of researchers in the United States, the United Kingdom, and France published their findings in *Nature* on May 13, 1961.[1] Both groups independently compared bacteriophage-infected and uninfected bacteria to track the formation of mRNA using radioactive nucleotides as labels. They postulated that the molecule had a role as an intermediate carrier of information from genes to ribosomes for protein synthesis.[2,3]

This groundbreaking discovery helped set the stage for the decades-long research conducted by many scientists to develop mRNA technologies. As the COVID-19 pandemic spread, these advances in mRNA therapeutics were applied to develop safe and effective vaccines against the coronavirus. One mRNA pioneer, Dr. Katalin Karikó, a biochemist and researcher at the University of Pennsylvania, graciously responded to my email inquiry about her work. "In science we get the knowledge and idea from those who came before us," she wrote.[4]

In 1978, Giorgos J. Dimitriadis, a Greek molecular biologist and biochemist, published a paper in *Nature* discussing how he introduced mRNA from one species, in this case rabbits, using liposomes (fat globules), into mouse lymphocytes to make proteins.[5] This seminal paper indicated that mRNA could be introduced into cells, which, in turn, could use it as a template to make proteins. Six years later, a paper describing how to make synthetic mRNA was published.[6] From these two papers, Karikó realized that mRNA could be used as a therapy by inserting it into cells and turning them into curative protein-making factories.

In the 1980s, Karikó wasn't thinking of using mRNA as a vaccine, although research papers published in 1993 and 1994 suggested that it might be possible.[7,8] She hypothesized that the process of inserting exogenous mRNA into cells might work in making proteins to treat diseases such as heart disease and strokes. She worked on this problem for ten years before discovering that the mRNA she was introducing was causing an inflammatory immune reaction. She had to learn why and figure out how to make it stop.

In 1997, Karikó met Dr. Drew Weissman, a recently hired assistant professor at Penn Medicine who had been working on HIV vaccines. They began a collaboration, culminating in a breakthrough seven years later, when they discovered that transfer RNA does not trigger an immune response. Understanding why required examining the interactions between DNA, RNA, and the immune system.

The immune system has two components: an older, innate component and a newer, adaptive component. The older, innate component identifies foreign substances such as invading microbes, but most importantly, it identifies self versus non-self. It has a system of proteins called toll-like receptors, located in the membranes of sentinel white blood cells, that bind to foreign proteins. If these toll-like receptors get activated, they set off a chain reaction, causing an inflammatory immune response.[9]

To avoid mistakenly attacking itself, the mammalian innate immune system modifies its DNA and RNA nucleotides with small molecules, a process that's analogous to placing a cloth over an object to hide it. By modifying its nucleotides, in effect "hiding" them, the body tells the toll-like receptors that the nucleotides are not foreign, but rather "self."[10] Karikó and her colleagues used different chemicals to modify their synthetic mRNA. The most useful chemical turned out to be pseudouridine, which proved to be very effective at diminishing the mRNA's in vivo immunogenicity. In addition, the modification improved the mRNA's stability and translational capability.[11]

With the promise of mRNA therapy revived, in 2013, Karikó joined BioNTech, a biotechnology company specializing in mRNA pharmaceuticals, cofounded by Drs. Ugur Sahin and Ozlem Tureci, a Turkish husband and wife medical immunologist-oncologist team, and Dr. Christopher Huber, chairman of Hematology and Oncology at the Johannes Gutenberg University of Mainz in Germany. In 2018, the company partnered with the pharmaceutical corporation Pfizer to develop an mRNA vaccine against influenza. As clinical trials were set to begin, the COVID-19 pandemic emerged in China. The BioNTech-Pfizer team realized that they needed to pivot and develop a COVID-19 vaccine as soon as possible. The mRNA technology facilitated the rapid development and manufacture of a highly effective and safe vaccine against the novel coronavirus.[12]

Over the decades, many researchers worked on mRNA as a potential therapeutic agent. At the Salk Institute for Biological Studies in La Jolla, California, a graduate student named Robert Malone performed a landmark study in the late 1980s: mixing droplets of lipids with strands of mRNA, which became absorbed by human cells and subsequently translated into proteins.[13]

Commercial interest in mRNA grew. Before BioNTech, other companies, including Merix Biosciences (later known as CoImmune), CureVac, and Moderna, were founded to work on mRNA therapeutics. In the early 2010s, the US Defense Advanced Research Projects Agency began funding mRNA vaccine research. When the COVID-19 pandemic began ravaging the world's populations, an infrastructure was in place to develop and manufacture mRNA vaccines. Within days of receiving the virus's genome, Moderna, the other company working on mRNA vaccines, collaborated with the National Institute of Allergy and Infectious Diseases to develop its vaccine.[14] Phase 1 studies in older adults showed that the vaccine caused mild or moderate side effects and induced neutralizing antibody titers after two injections administered twenty-eight days apart, producing antibody levels similar to those in individuals who had recovered from COVID-19 infection.[15]

On December 11, 2020, the US Food and Drug Administration (FDA) approved the BioNTech-Pfizer COVID-19 vaccine under Emergency Use Authorization (EUA) for individuals 16 years of age and older. It was the first mRNA vaccine approved to prevent COVID-19.[16] One week later, the FDA issued an EUA for Moderna's mRNA vaccine.[17] The challenge, however, was to convince the public that the novel vaccines were safe and efficacious.

Notably, gain-of-function research was not needed for the development of the COVID-19 vaccines.[18]

Notes

Preface

1. Kahn LH. Confronting zoonoses, linking human and veterinary medicine. *Emerging Infectious Diseases*. 2016;12(4):556–561. doi:10.3201/eid1204.050956.

2. Saunders LZ. Virchow's contributions to veterinary medicine: celebrated then, forgotten now. *Veterinary Pathology*. 2000;37(3):199–207. doi:10.1354/vp.37-3-199; Saunders LZ. From Osler to Olafson: the evolution of veterinary pathology in North America. *Canadian Journal of Veterinary Research*. 1987;51(1): 1–26. https://www.ncbi.nlm.nih.gov/pmc/articles/PMC1255268/pdf/cjvetres00057-0003.pdf. Accessed January 16, 2023; Correction. *Canadian Journal of Veterinary Research*. 1987;51(2): 284. https://www.ncbi.nlm.nih.gov/pmc/articles/PMC1255268/. Accessed January 16, 2023.

3. Kahn LH. Going viral. *Bulletin of the Atomic Scientists*. January 17, 2012. https://thebulletin.org/2012/01/going-viral/#post-heading. Accessed January 21, 2021.

Introduction

1. According to the National Institute of Allergy and Infectious Diseases (of the US National Institutes of Health), an emerging infectious disease is defined as an infectious disease that has newly appeared in a population, or one that has existed but is rapidly increasing in incidence or geographic range, or one that is caused by one of the NIAID Category A, B, or C priority pathogens. https://www.niaid.nih.gov/research/emerging-infectious-diseases-pathogens. Accessed January 15, 2023.

2. Düx A, Lequime S, Patrono LV, et al. Measles virus and rinderpest virus diverged dated to the sixth century BCE. *Science*. 2020;368(6497):1367–1370. doi:10.1126/science/aba9411.

3. The rinderpest virus is the second virus, after smallpox, to be eradicated. Vaccines against smallpox and rinderpest were essential for the successful eradication efforts.

4. Zhou Z, Qiu Y, Ge X. The taxonomy, host range and pathogenicity of coronaviruses and other viruses in the Nidovirales order. *Animal Diseases*. 2021;1(5). doi:10.1186/s44149-021-00005-9.

5. Duffy S. Why are RNA virus mutation rates so damn high? *PLoS Biology*. 2018;16(8). doi:10.1371/journal.pbio.3000003.

6. Jimenez J. Can you find eggs here or there? Can you find them anywhere? *New York Times*. January 12, 2023. https://www.nytimes.com/2023/01/12/us/egg-shortage-us.html. Accessed January 18, 2023.

7. Stokstad E. Wrestling with bird flu, Europe considers once-taboo vaccines. *Science*. May 11, 2022. https://www.science.org/content/article/wrestling-bird-flu-europe-considers-once-taboo-vaccines. Accessed January 22, 2023.

8. According to the Association for Professionals in Infection Control and Epidemiology, an outbreak is generally defined as a sudden rise in the number of cases of a disease, typically in a community or well-defined geographic area. An epidemic is defined as an infectious disease spreading rapidly to many people, and a pandemic is a global disease outbreak, affecting many more people than an epidemic. https://apic.org/monthly_alerts/outbreaks-epidemics-and-pandemics-what-you-need-to-know/. Accessed January 15, 2023.

9. US Centers for Disease Control and Prevention. About MERS. https://www.cdc.gov/coronavirus/mers/about/index.html. Accessed March 17, 2022.

10. Report of the WHO Consultative Meeting on High/Maximum Containment (Biosafety Level 4) Laboratories Networking, Lyon, France. 13–15 December 2017. Archived February 8, 2021 on the Wayback Machine. Geneva: World Health Organization; 2018 (WHO/WHE/CPI/2018.40). https://apps.who.int/iris/bitstream/handle/10665/311625/WHO-WHE-CPI-2018.40-eng.pdf?sequence=1). Accessed October 26, 2021.

11. Cyranoski D. Inside the Chinese lab poised to study world's most dangerous pathogens. *Nature*. 2017;542:399–400. doi:10.1038/nature.2017.21487.

12. Parry J. Breaches of safety regulations are probably cause of recent SARS outbreak, WHO says. *British Medical Journal*. 2004;328(7450):1222. doi:10.1136/bmj.328.7450.1222-b.

13. Klotz L. Human error in high-biocontainment labs: a likely pandemic threat. *Bulletin of the Atomic Scientists*. February 25, 2019. https://thebulletin.org/2019/02/human-error-in-high-biocontainment-labs-a-likely-pandemic-threat/. Accessed October 26, 2021.

14. Calisher C, Carroll D, Colwell R, et al. Statement in support of the scientists, public health professionals, and medical professionals of China combatting COVID-19. *Lancet*. 2020;395(10226):E42–E43. doi:10.1016/S0140-6736(20)30418-9.

15. Anderson KG, Rambaut A, Lipkin WI, et al. The proximal origin of SARS-CoV-2. *Nature Medicine*. 2020;26:450–452. doi:10.1038/s41591-020-0820-9.

16. Relman DA. Opinion: to stop the next pandemic, we need to unravel the origins of COVID-19. *Proceedings of the National Academy of Science*. 2020;117(47):29246–29248. doi:10.1073/pnas.2021133117.

17. Rogers K, Jakes L, Swanson A. Trump defends using "Chinese virus" label, ignoring growing criticism." *New York Times*. March 18, 2020. https://www.nytimes.com/2020/03/18/us/politics/china-virus.html. Accessed October 27, 2021; Nakamura D. With "kung flu," Trump sparks backlash over racist language—and a rallying cry for supporters. *Washington Post*. June 24, 2020. https://www.washingtonpost.com/politics/with-kung-flu-trump-sparks-backlash-over-racist-language--and-a-rallying-cry-for-supporters/2020/06/24/485d151e-b620-11ea-aca5-ebb63d27e1ff_story.html. Accessed October 27, 2021.

18. Rogers K. Politicians' use of "Wuhan virus" starts a debate health experts wanted to avoid. *New York Times*. March 10, 2020. https://www.nytimes.com/2020/03/10/us/politics/wuhan-virus.html. Accessed October 27, 2021.

19. Tavernise S, Oppel RA Jr. Spit on, yelled at, attacked: Chinese-Americans fear for their safety. *New York Times*. March 23, 2020. https://www.nytimes.com/2020/03/23/us/chinese-coronavirus-racist-attacks.html. Accessed: October 26, 2021.

20. Farhi P, Barr J. The media called the "lab leak" story a "conspiracy theory." Now it's prompted corrections—and serious new reporting. *Washington Post*. June 10, 2021. https://www.washingtonpost.com/lifestyle/media/the-media-called-the-lab-leak-story-a-conspiracy-theory-now-its-prompted-corrections--and-serious-new-reporting/2021/06/10/c93972e6-c7b2-11eb-a11b-6c6191ccd599_story.html. Accessed October 27, 2021.

21. Zimmer C, Mueller B. New research points to Wuhan market as pandemic origin. *New York Times*. February 27, 2022. https://www.nytimes.com/interactive/2022/02/26/science/covid-virus-wuhan-origins.html?smid=em-share. Accessed February 28, 2022.

22. Worobey M, Levy JI, Malpica LM, et al. The Huanan market was the epicenter of SARS-CoV-2 emergence. Zenodo. 2022. doi:10.5281/zenodo.6299116.

23. Pekar JE, Magee A, Parker E, et al. SARS-CoV-2 emergence very likely resulted from at least two zoonotic events. Zenodo. 2022. doi:10.5281/zenodo.6291628.

24. Zimmer C, Mueller B, Buckley C. First known covid case was vendor at Wuhan market, scientist says. *New York Times*. November 18, 2021. https://www.nytimes.com/2021/11/18/health/covid-wuhan-market-lab-leak.html. Accessed February 28, 2022.

25. Eban K. The lab-leak theory: inside the fight to uncover COVID-19's origins. *Vanity Fair*. June 3, 2021. https://www.vanityfair.com/news/2021/06/the-lab-leak-theory-inside-the-fight-to-uncover-covid-19s-origins. Accessed November 10, 2021.

26. Gao G, Liu W, Liu P, et al. Surveillance of SARS-CoV-2 in the environment and animal samples of the Huanan Seafood Market. *Nature Portfolio*. 2022. doi:10.21203/rs.3.rs-1370392/v1.

27. Kimmerer RW. Weaving traditional ecological knowledge into biological education: a call to action. *BioScience*. 2002;52(5):432–438. doi:10.1641/0006-3568(2002)052[0432:WTEKIB]2.0.CO;2.

28. This statement excludes the national surveillance systems of insect plant pests and pollinators, which are part of the animal kingdom.

29. Rabinowitz PM, Natterson-Horowitz BJ, Kahn LH, et al. Incorporating one health into medical education. *BMC Medical Education*. 2017;17:45. doi:10.1186/s12909-017-0883-6; Kahn LH. Developing a one health approach by using a multi-dimensional matrix. *One Health*. 2021;13:100289. doi:10.1016/j.onehlt.2021.100289. Intersecting multidimensions is technically called a tensor, but in this book, we will call it a matrix.

CHAPTER ONE: Domestic and Wild Animals

1. Schalk AF, Hawn MC. An apparently new respiratory disease of baby chicks. Proceedings: Thirty-fourth Annual Meeting of the United States Live Stock Sanitary Association. Chicago, Illinois. December 1930:413–423. https://www.usaha.org/upload/Proceedings/1930-1959/1930_THIRTY_FOURTH_ANNUAL_MEETIN.pdf. Accessed October 1, 2020; Schalk AF. An apparently new respiratory disease of baby chicks. *Journal of the American Veterinary Medical Association*. 1931;78:413–423.

2. Lalchhandama K. A biography of coronaviruses from IBV to SARS-CoV-2, with their evolutionary paradigms and pharmacological challenges. *International Journal of Research in Pharmaceutical Sciences*. 2020;11(special issue 1):208–218. doi:10.26452/ijrps.v11iSPL1.2701.

3. Bushnell LD, Brandly CA. Laryngotracheitis in chicks. *Poultry Science.* 1933;12(1):55–60. doi:10.3382/ps0120055. Accessed October 2, 2020.

4. Dahl JL, Gatlin W. A microbiology teaching lab: using Koch's postulates to determine the cause of "Peep pox" in marshmallow Peeps. *American Biology Teacher.* 2018;80(9):676–679. doi:10.1525/abt2018.80.9.676. Koch's postulates involve four steps to prove that a microbe causes disease: (1) The microbe must be present in all instances of individuals with the disease; (2) The microbe must be isolated in pure culture; (3) A test animal infected with the purified microbe will develop signs resembling those of the original host; (4) The same microbe must be isolated from the test animal showing similar signs.

5. Artenstein AW. The discovery of viruses: advancing science and medicine by challenging dogma. *International Journal of Infectious Diseases.* 2012;16(7):e470–e473. doi:10.1016/j.ijid.2012.03.005.

6. Rockefeller University. Discovering that viruses are made up of protein and nucleic acid. http://centennial.rucares.org/index.php?page=Protein_Nucleic_Acid. Accessed December 25, 2020. For this work, Wendell M. Stanley was awarded the Nobel Prize in 1946. Nobel Prize. William M. Stanley biographical. https://www.nobelprize.org/prizes/chemistry/1946/stanley/biographical/. Accessed December 25, 2020.

7. Machemer T. How a few sick tobacco plants led scientists to unravel the truth about viruses. *Smithsonian Magazine.* March 24, 2020. https://www.smithsonianmag.com/science-nature/what-are-viruses-history-tobacco-mosaic-disease-180974480/. Accessed October 6, 2020.

8. Beach JR. The virus of laryngotracheitis of fowls. *Science.* 1930;72(1877):633–634. doi:10.1126/science.72.1877.633.

9. Lalchhandama K. A biography of coronaviruses from IBV to SARS-CoV-2, with their evolutionary paradigms and pharmacological challenges. *International Journal of Research Pharmaceutical Sciences.* 2020;11(special issue 1):208–218. doi:10.26452/ijrps.v11iSPL1.2701.

10. Beach JR, Schalm OW. A filterable virus, distinct from that of laryngotracheitis, the cause of a respiratory disease of chicks. *Poultry Science.* 1936;15(3):199–206. doi:10.3382/ps.0150199.

11. Hinshaw WR. Fred Robert Beaudette: distinguished avian microbiologist and pathologist 1897–1958. *Avian Diseases.* 1957;1(1):2–17. http://www.jstor.org/stable/1587540; Hudson CB. Nunc Dimittis. *Poultry Science* 1969;48(3):1137. doi:10.3382/ps.0481137.

12. Beaudette FR, Hudson CB. Cultivation of the virus of infectious bronchitis. *JAVMA.* 1937;90:51–58.

13. Goldsmith CS, Miller SE. Modern uses of electron microscopy for detection of viruses. *Clinical Microbiology Reviews.* 2009;22(4):552–563. doi:10.1128/CMR.00027-09; Kausche G, Pfankuch E, Ruska H. Die Sichtbarmachung von pflanzlichem Virus im Übermikroskop. *Naturwissenschaften.* 1939;27:292–299.

14. Nobel Prize Organization. Ernst Ruska biographical. https://www.nobelprize.org/prizes/physics/1986/ruska/biographical/. Accessed October 30, 2020.

15. Berry DM, Cruickshank JG, Chu HP, et al. The structure of infectious bronchitis virus. *Virology.* 1964;23:403–407. doi:10.1016/0042-6822(64)90263-6.

16. Virology: coronaviruses. *Nature.* 1968;220:650. doi:10.1038/220650b0.

17. World Health Organization. Summary of probable SARS cases with onset of illness from 1 Nov 2002 to 31 July 2003. https://www.who.int/publications/m

/item/summary-of-probable-sars-cases-with-onset-of-illness-from-1-november-2002-to-31-july-2003. Accessed December 9, 2023.

18. McKay B, Dvorak P. A deadly coronavirus was inevitable. Why was no one ready? *Wall Street Journal*. August 13, 2020. https://www.wsj.com/articles/a-deadly-coronavirus-was-inevitable-why-was-no-one-ready-for-covid-11597325213. Accessed December 26, 2020.

19. Xu R-H, He J-F, Evans MR, et al. Epidemiologic clues to SARS origin in China. *Emerging Infectious Diseases*. 2004;10(6):1030–1037. doi:10.3201/eid1006.030852.

20. Guan Y, Zheng BJ, He YQ, et al. Isolation and characterization of viruses related to the SARS coronavirus from animals in southern China. *Science*. 2003;302:276–278. doi:10.1126/science.1087139.

21. Tu C, Crameri G, Kong X, et al. Antibodies to SARS-coronavirus in civets. *Emerging Infectious Diseases*. 2004;10(12):2244–2248. doi:10.3201/eid1012.040520.

22. Shi Z, Hu Z. A review of studies on animal reservoirs of the SARS coronavirus. *Virus Research*. 2008:133(1):74–87. doi:10.1016/j.virusres.2007.03.012.

23. Koerner B. What does civet cat taste like? *Slate Magazine*. January 6, 2004. https://slate.com/news-and-politics/2004/01/what-does-civet-cat-taste-like.html. Accessed January 24, 2021.

24. Linfa Wang. Professor and director, Emerging Infectious Diseases Program. Duke Global Health Institute. Duke University. Curriculum vitae. https://globalhealth.duke.edu/people/wang-linfa. Accessed December 28, 2020.

25. Dimkic I, Fira D, Janakiev T, et al. The microbiome of bat guano: for what is this knowledge important? *Applied Microbiology and Biotechnology*. 2021:105:1407–1419. doi:10.1007/s00253-021-11143-y.

26. Wendong L, Zhengli S, Meng Y, et al. Bats are natural reservoirs of SARS-like coronaviruses. *Science*. 2005;310(5748):676–679. doi:10.1126/science.1118391.

27. Lau SKP, Woo PCY, Li KSM, et al. Severe acute respiratory syndrome coronavirus-like virus in Chinese horseshoe bats. *Proceedings of the National Academy of Science*. 2005;102(39):14040–14045. doi:10.1073/pnas.0506735102.

28. Li W, Shi Z, Yu M, et al. Bats are the natural reservoirs of SARS-like coronaviruses. *Science*. 2005;310(5748):676–679. doi:10.1126/science.1118391.

29. Shi Z, Hu Z. A review of studies on animal reservoirs of the SARS coronavirus. *Virus Research*. 2008;133(1):74–87. doi:10.1016/j.virusres.2007.03.012.

30. Shi Z, Hu Z. A review of studies on animal reservoirs of the SARS coronavirus. *Virus Research*. 2008;133(1):74–87. doi:10.1016/j.virusres.2007.03.012.

31. Teeling EC, Springer MS, Madsen O, et al. A molecular phylogeny for bats illuminates biogeography and the fossil record. *Science*. 2005;307(5709):580–584. doi:10.1126/science.1105113.

32. S.H. Newman, H.E. Field, C.E. de Jong, J.H. Epstein, eds. *Investigating the role of bats in emerging zoonoses: balancing ecology, conservation and public health interests*. Rome: Food and Agriculture Organization of the United Nations; 2011. http://www.fao.org/3/a-i2407e.pdf. Accessed December 30, 2020.

33. O'Shea TJ, Cryan PM, Cunningham AA, et al. Bat flight and zoonotic viruses. *Emerging Infectious Diseases*. 2014;20(5):741–745. doi:10.3201/eid2005.130539.

34. Banerjee A, Baker ML, Kulcsar K, Misra V, Plowright R, Mossman K. Novel insights into immune systems of bats. *Frontiers in Immunology*. 2020;11. doi:10.3389/fimmu.2020.00026.

35. Ahn M, Anderson DE, Zhang Q, et al. Dampened NLRP3-mediated inflammation in bats and implications for a special viral reservoir host. *Nature Microbiology.* 2019;4:789–799. doi:10.1038/s41564-019-0371-3.

36. Calisher CH, Childs JE, Field HE, et al. Bats: important reservoir hosts of emerging viruses. *Clinical Microbiology Reviews.* 2006;19(3):531–545. doi:10.1128/CMR.00017-06.

37. Chen L, Liu B, Yang J, et al. DBatVir: the database of bat-associated viruses. *Database.* 2014. doi:10.1093/database/bau021.

38. Database of bat-associated viruses. http://www.mgc.ac.cn/DBatVir/. Accessed December 31, 2020.

39. Lu L, Zhong W, Bian Z, et al. A comparison of mortality-related risk factors of COVID-19, SARS, and MERS: a systematic review and meta-analysis. *Journal of Infection.* 2020;81(4):e18–e25. doi:10.1016/j.jinf.2020.07.002.

40. Zaki AM, van Boheemen S, Bestebroer TM, et al. Isolation of a novel coronavirus from a man with pneumonia in Saudi Arabia. *New England Journal of Medicine.* 2012;367(19):1814–1820. doi:10.1056/NEJMoa1211721.

41. Mohd HA, Al-Tawfiq JA, Memish ZA. Middle East respiratory syndrome coronavirus (MERS-CoV) origin and animal reservoir. *Virology Journal.* 2016;13:87. doi:10.1186/s12985-016-0544-0.

42. Hemida MG, Elmoslemany A, Al-Hizab F, et al. Dromedary camels and the transmission of Middle East respiratory syndrome coronavirus (MERS-CoV). *Transboundary and Emerging Diseases.* 2017;64(2):344–353. doi:10.1111/tbed.12401.

43. Alagaili AN, Briese T, Mishra N, et al. Middle East respiratory syndrome coronavirus infection in dromedary camels in Saudi Arabia. *mBio.* 2014;5(2):e00884-14. doi:10.1128/mBio.00884-14.

44. Hemida MG, Chu DKW, Poon LLM, et al. MERS coronavirus in dromedary camel herd, Saudi Arabia. *Emerging Infectious Diseases.* 2014;20(7):1231–1234. doi:10.3201/eid2007.140571.

45. Orlando L. Back to the roots and routes of dromedary domestication. *Proceedings of the National Academy of Science.* 2016;113(24):6588–6590. doi:10.1073/pnas.1606340113.

46. Al Abri MA, Faye B. Genetic improvement in dromedary camels: challenges and opportunities. *Frontiers in Genetics.* 2019;10. doi:10.3389/fgene.2019.00167.

47. Shi Z and Hu Z. A review of studies on animal reservoirs of the SARS coronavirus. *Virus Research.* 2008;133(1):74–87. doi:10.1016/j.virusres.2007.03.012.

48. Killerby ME, Biggs HM, Midgley CM, et al. Middle East respiratory syndrome coronavirus transmission. *Emerging Infectious Diseases.* 2020;26(2):191–198. doi:10.3201/eid2602.190697.

49. Memish ZA, Mishra N, Olival KJ, et al. Middle East respiratory syndrome coronavirus in bats, Saudi Arabia. *Emerging Infectious Diseases.* 2013;19(11). doi:10.3201/eid1911.131172.

50. Kupferschmidt K. Bat out of hell? Egyptian tomb bat may harbor MERS virus. *Science.* August 22, 2013. https://www.sciencemag.org/news/2013/08/bat-out-hell-egyptian-tomb-bat-may-harbor-mers-virus. Accessed January 21, 2021.

51. Ithete N, Stoffberg S, Corman V, et al. Close relative of human Middle East respiratory syndrome coronavirus in bat, South Africa. *Emerging Infectious Diseases.* 2013;19(10):1697–1699. doi:10.3201/eid1910.130946.

52. Annan A, Baldwin HJ, Corman V, et al. Human betacoronavirus 2c EMC/2012-related viruses in bats, Ghana and Europe. *Emerging Infectious Diseases*. 2013;19(3):456–459. doi:10.3201/eid1903.121503.

53. Kindler E, Jonsdottir HR, Muth D, et al. Efficient replication of the novel human betacoronavirus EMC on primary human epithelium highlights its zoonotic potential. *mBio*. 2013;4(1):e00611-12. doi:10.1128/mBio.00611-12.

54. Sharif-Yakan A, Kanj SS. Emergence of MERS-CoV in the Middle East: origins, transmission, treatment, and perspectives. *PLoS Pathogens*. 2014;10(12):e1004457. doi:10.1371/journal.ppat.1004457.

55. Haagmans BL, Al Dhahiry SHS, Reusken CBEM, et al. Middle East respiratory syndrome coronavirus in dromedary camels: an outbreak investigation. *Lancet Infectious Diseases*. 2014;14(2):P140–145. doi:10.1016/S1473-3099(13)70690-X.

56. Killerby ME, Biggs HM, Midgley CM, et al. Middle East respiratory syndrome coronavirus transmission. *Emerging Infectious Diseases*. 2020;26(2):191–198. doi:10.3201/eid2602.190697.

57. ProMED-mail. Undiagnosed pneumonia—China (Hubei). Request for information. ProMED-mail 2019; 30 Dec: Archive Number 20191230.6864153. https://promedmail.org/promed-post/?id=6864153. Accessed January 31, 2021.

58. World Health Organization. Origin of SARS-CoV-2. March 26, 2020. https://apps.who.int/iris/bitstream/handle/10665/332197/WHO-2019-nCoV-FAQ-Virus_origin-2020.1-eng.pdf. Accessed January 31, 2021.

59. Page J. "Virus sparks soul-searching over China's wild animal trade; Beijing faces uncomfortable questions over its failure to clean up wildlife trade and public calls for a permanent ban on wild meat." *Wall Street Journal*. January 26, 2020. https://www.wsj.com/articles/virus-sparks-soul-searching-over-chinas-wild-animal-trade-11580055290. Accessed March 17, 2020.

60. China Global Television Network (CGTN). Official: Wuhan seafood market may be the victim of the coronavirus. May 26, 2020. https://news.cgtn.com/news/2020-05-26/Official-Wuhan-seafood-market-may-be-the-victim-of-the-coronavirus--QNVxtMlFbW/index.html. Accessed January 31, 2021.

61. Gao G, Liu W, Liu P, et al. Surveillance of SARS-CoV-2 in the environment and animal samples of the Huanan seafood market. *Nature Portfolio*. 2022. doi:10.21203/rs.3.rs-1370392/v1; Liu WJ, Liu P, Lei W, et al. Surveillance of SARS-CoV-2 at the Huanan Seafood Market. *Nature*. 2023. doi:10.1038/s41586-023-06043-2.

62. Huang C, Wang Y, Li X, et al. Clinical features of patients infected with 2019 novel coronavirus in Wuhan China. *Lancet*. 2020;395(10223):P497–506. doi:10.1016/S0140-6736(20) 30183–5.

63. Cohen J. Wuhan seafood market may not be source of novel virus spreading globally. *Science*. January 26, 2020. https://www.sciencemag.org/news/2020/01/wuhan-seafood-market-may-not-be-source-novel-virus-spreading-globally. Accessed January 31, 2021.

64. Kuman S, Tao Q, Weaver S, et al. An evolutionary portrait of the progenitor SARS-CoV-2 and its dominant offshoots in COVID-19 pandemic. *bioRxiv*. 2020. doi:10.1101/2020.09.24.311845.

65. Ge X-Y, Yang W-H, Zhou J-H, et al. Detection of alpha- and betacoronaviruses in rodents from Yunnan, China. *Virology Journal*. 2017;14(98). doi:10.1186/s12985-017-0766–9.

66. Zhou P, Yang X-L, Wang X-G, et al. A pneumonia outbreak associated with a new coronavirus of probable bat origin. *Nature*. 2020;579(7798):270–273. doi:10.1038/s41586-020-2012-7.

67. Ge X-Y, Wang N, Zhang W, et al. Coexistence of multiple coronaviruses in several bat colonies in an abandoned mineshaft. *Virologica Sinica*. 2016;31(1):31–40. doi:10.1007/s12250-016-3713-9.

68. Quammen D. "Did pangolin trafficking cause the coronavirus pandemic?" *New Yorker.* August 24, 2020. https://www.newyorker.com/magazine/2020/08/31/did-pangolins-start-the-coronavirus-pandemic. Accessed February 4, 2021.

69. Ge X-Y, Li J-L, Yang X-L, et al. Isolation and characterization of a bat SARS-like coronavirus that uses the ACE2 receptor. *Nature*. 2013;503:535–538. doi:10.1038/nature12711.

70. Chinese medicine and the pangolin. *Nature*. 1938;141:72. doi:10.1038/141072b0.

71. Volpato G, Fontefrancesco MF, Gruppuso P, et al. Baby pangolins on my plate: possible lessons to learn from the COVID-19 pandemic. *Journal of Ethnobiology and Ethnomedicine*. 2020;16:19. doi:10.1186/s13002-020-00366-4.

72. Convention on Illegal Trade in Endangered Species of Wild Fauna and Flora. Checklist of CITES species. https://checklist.cites.org/#/en/search/output_layout=alphabetical&level_of_listing=0&show_synonyms=1&show_author=1&show_english=1&show_spanish=1&show_french=1&scientific_name=Pangolin&page=1&per_page=20. Accessed February 5, 2021.

73. Lam TTY, Jia N, Zhang YW, et al. Identifying SARS-CoV-2-related coronaviruses in Malayan pangolins. *Nature*. 2020;583:282–285. doi:10.1038/s41586-020-2169-0.

74. Liu P, Jiang J-Z, Wan X-F, et al. Are pangolins the intermediate host of the 2019 novel coronavirus (SARS-CoV-2)? *PLOS Pathogens* 2020;16(5):e1008421. doi:10.1371/journal.ppat.1008421.

75. Xiao K, Zhai J, Feng Y, et al. Isolation and characterization of 2019-nCoV-like coronavirus from Malayan pangolins. *bioRxiv*. 2020. doi:10.1101/2020.02.17.951335.

76. Callaway E, Cyranoski D. Why snakes probably aren't spreading the new China virus. *Nature*. January 23, 2020. doi:10.1038/d41586-020-00180-8.

77. Wang W, Tian J-H, Chen X, et al. Coronaviruses in wild animals sampled in and around Wuhan in the beginning of COVID-19 emergence. *Virus Evolution*. 2022;8(1):veac046. doi:10.1093/ve/veac046.

78. CDC. Animals and COVID-19. https://www.cdc.gov/coronavirus/2019-ncov/daily-life-coping/animals.html?CDC_AA_refVal=https%3A%2F%2Fwww.cdc.gov%2Fcoronavirus%2F2019-ncov%2Fanimals%2Fpets-other-animals.html. Accessed January 8, 2022.

79. Leroy EM, Gouilh MA, Brugere-Picoux J. The risk of SARS-CoV-2 transmission to pets and other wild and domestic animals strongly mandates a one-health strategy to control the COVID-19 pandemic. *One Health*. 2020;10:100133. doi:10.1016/j.onehlt.2020.100133.

80. US Department of Agriculture. Animal and plant health inspection service. Questions and answers: results of study on SARS-CoV-2 in white-tailed deer. August 2021. https://www.aphis.usda.gov/animal_health/one_health/downloads/qa-covid-white-tailed-deer-study.pdf. Accessed January 8, 2022.

81. Gorman J. Mink and the coronavirus: what we know. *New York Times*. December 23, 2020. https://www.nytimes.com/article/mink-coronavirus-mutation.html. Accessed February 11, 2021.

82. Munnink BBO, Sikkema RS, Nieuwenhuijse DF, et al. Transmission of SARS-CoV-2 on mink farms between humans and mink and back to humans. *Science*. 2021;371(6525):172–177. doi:10.1126/science.abe5901.

83. Mallapaty S. "COVID mink analysis shows mutations are not dangerous—yet." *Nature*. November 16, 2020. https://www.nature.com/articles/d41586-020-03218-z. Accessed February 11, 2021.

84. Sorensen MS. After Denmark's mink cull, questions over legality, science, and what to do with "zombie minks." *Washington Post*. December 18, 2020. https://www.washingtonpost.com/world/europe/denmark-mink-coronavirus-zombie-/2020/12/17/1f8cb00e–3bea-11eb-aad9-8959227280c4_story.html. Accessed February 11, 2021.

85. Wei C, Shan K-J, Wang W, et al. Evidence for a mouse origin of the SARS-CoV-2 Omicron variant. *Journal of Genetics and Genomics*. 2021;48(12)1111–1121. doi:10.1016/j.jgg.2021.12.003.

CHAPTER TWO: Environments and Ecosystems

1. Suttle CA. Viruses in the sea. *Nature*. 2005;437:356–361. doi:10.1038/nature04160.

2. Suttle CA. Marine viruses—major players in the global ecosystem. *Nature Reviews Microbiology*. 2007;5:801–812. doi:10.1038/nrmicro1750.

3. Hall JS, Dusek RJ, Nashold SW, et al. Avian influenza virus prevalence in marine birds is dependent on ocean temperatures. *Ecological Applications*. 2020;30(2):e02040. doi:10.1002/eap.2040.

4. University of British Columbia. Viruses—lots of them—are falling from the sky: an astonishing number of viruses are circulating around the Earth's atmosphere—and falling from it. *ScienceDaily*, February 6, 2018. www.sciencedaily.com/releases/2018/02/180206090650.htm.

5. Reche I, D'Orta G, Mladenov N, et al. Deposition rates of viruses and bacteria above the atmospheric boundary layer. *ISME Journal*. 2018;12:1154–1162. doi:10.1038/s41396-017-0042-4.

6. Sharoni S, Trainic M, Schatz D, et al. Infection of phytoplankton by aerosolized marine viruses. *Proceedings of the National Academy of Science*. 2015;112(21):6643–6647. doi:10.1073/pnas.1423667112.

7. USAID. Emerging pandemic threats. https://www.usaid.gov/news-information/fact-sheets/emerging-pandemic-threats-program. Accessed June 15, 2021.

8. Kelly TR, Machalaba C, Karesh WB, et al. Implementing One Health approaches to confront emerging and re-emerging zoonotic disease threats: lessons from PREDICT. *One Health Outlook*. 2020;2. doi:10.1186/s42522-019-0007-9.

9. Anthony SJ, Johnson CK, Greig DJ, et al. Global patterns in coronavirus diversity. *Virus Evolution*. 2017;3(1):vex012. doi:10.1093/ve/vex012.

10. Centers for Disease Control and Prevention. Scientific brief: SARS-CoV-2 transmission. Summary of recent changes. https://www.cdc.gov/coronavirus/2019-ncov/science/science-briefs/sars-cov-2-transmission.html#anchor_1619805240227. Accessed July 12, 2021.

11. Pan M, Lednicky JA, Wu C-Y. Collection, particle sizing and detection of airborne viruses. *Journal of Applied Microbiology.* 2019;127(6):1596–1611. doi:10.1111/jam.14278.

12. Jones DL, Baluja MQ, Graham DW, et al. Shedding of SARS-CoV-2 in feces and urine and its potential role in person-to-person transmission and the environment-based spread of COVID-19. *Science of the Total Environment.* 2020;749:141364. doi:10.1016/j.scitotenv.2020.141364.

13. Lewis D. Superspreading drives the COVID pandemic—and could help to tame it. *Nature.* February 23, 2021. https://www.nature.com/articles/d41586-021-00460-x. Accessed July 12, 2021.

14. Kraay ANM, Hayashi MAL, Berendes DM, et al. Risk for fomite-mediated transmission of SARS-CoV-2 in child daycares, schools, nursing homes, and offices. *Emerging Infectious Diseases.* 2021;27(4):1229–1231. doi:10.3201/eid2704.203631.

15. Leung NHL. Transmissibility and transmission of respiratory viruses. *Nature Reviews Microbiology.* 2021. doi:10.1038/s41579-021-00535-6.

16. Delamater PL, Street EJ, Leslie TF, et al. Complexity of the basic reproduction number (R_0) *Emerging Infectious Diseases.* 2019;25(1):1–4. doi:10.3201/eid2501.171901.

17. Guerra FM, Bolotin S, Lim G, et al. The basic reproduction number (R_0) of measles: a systematic review. *Lancet Infectious Diseases.* 2017;17(12):e420–e428. doi:10.1016/S1473-3099(17)30307-9.

18. Leung NHL. Transmissibility and transmission of respiratory viruses. *Nature Reviews Microbiology.* 2021. doi:10.1038/s41579-021-00535-6.

19. Department of Communicable Disease Surveillance and Response. World Health Organization. Consensus document on the epidemiology of severe acute respiratory syndrome (SARS). WHO/CDS/CSR/GAR/2003.11. WHO. May 16–17, 2003:9, 12. https://www.who.int/csr/sars/en/WHOconsensus.pdf. Accessed June 17, 2021.

20. Zimmer K. Why R_0 is problematic for predicting COVID-19 spread. *Scientist.* July 13, 2020. https://www.the-scientist.com/features/why-r0-is-problematic-for-predicting-covid-19-spread-67690. Accessed June 24, 2021.

21. Yu Its, Li Y, Wong TW, et al. Evidence of airborne transmission of the severe acute respiratory syndrome virus. *New England Journal of Medicine.* 2004;350:1731–1739. doi:10.1056/NEJMoa032867.

22. World Health Organization. Summary of probable SARS cases with onset of illness from 1 November 2002 to 31 July 2003. https://www.who.int/publications/m/item/summary-of-probable-sars-cases-with-onset-of-illness-from-1-november-2002-to-31-july-2003. Accessed July 4, 2021.

23. Bamford C. The original SARS virus disappeared—here's why coronavirus won't do the same. *Conversation.* June 5, 2020. https://theconversation.com/the-original-sars-virus-disappeared-heres-why-coronavirus-wont-do-the-same-138177. Accessed July 13, 2021.

24. Walgate R. SARS escaped Beijing lab twice. *Genome Biology.* 2004. doi:10.1186/gb-spotlight-20040427-03.

25. Parry J. Breaches of safety regulations are probable cause of recent SARS outbreak, WHO says. *British Medical Journal.* 2004;328(7450):1222. doi:10.1136/bmj.328.7450.1222-b.

26. Zimmer K. Why R_0 is problematic for predicting COVID-19 spread. *Scientist.* July 13, 2020. https://www.the-scientist.com/features/why-r0-is-problematic-for-predicting-covid-19-spread-67690. Accessed June 24, 2021.

27. Killerby ME, Biggs HM, Midgley CM, et al. Middle East respiratory syndrome coronavirus transmission. 2020;26(2):191–198. doi:10.3201/eid2602.190697.

28. Xiao S, Li Y, Sung M, et al. A study of the probable transmission routes of MERS-CoV during the first hospital outbreak in the Republic of Korea. *Indoor Air.* 2018;28:51–63. doi:10.1111/ina.12430.

29. Kim KM, Ki M, Cho S-il, et al. Epidemiologic features of the first MERS outbreak in Korea: focus on Pyeongtaek St. Mary's Hospital. *Epidemiology and Health.* 2015;37:e2015041. doi:10.4178/epih/e2015041.

30. Xiao S, Li Y, Sung M, et al. A study of the probable transmission routes of MERS-CoV during the first hospital outbreak in the Republic of Korea. *Indoor Air.* 2018;28:51–63. doi:10.1111/ina.12430.

31. Park J-M. Tracing the path of South Korea's MERS "patient zero." *Scientific American.* June 4, 2015. https://www.scientificamerican.com/article/tracing-the-path-of-south-korea-s-mers-patient-zero/. Accessed July 2, 2021.

32. Van Doremalen N, Bushmaker T, Munster VJ. Stability of Middle East respiratory syndrome coronavirus (MERS-CoV) under different environmental conditions. *Eurosurveillance.* 2013;18(38):20590. doi:10.2807/1560-7917.ES2013.18.38.20590.

33. Killerby ME, Biggs HM, Midgley CM, et al. Middle East respiratory syndrome coronavirus transmission. *Emerging Infectious Diseases.* 2020;26(2):191–198. doi:10.3201/eid2602.190697.

34. AlBalwi MA, Khan A, AlDrees M, et al. Evolving sequence mutations in the Middle East respiratory syndrome coronavirus (MERS-CoV). *Journal of Infection and Public Health.* 2020;13(10):1544–1550. doi:10.1016/j.jiph.2020.06.030.

35. Wei WE, Li Z, Chiew CJ, et al. Presymptomatic transmission of SARS-CoV-2—Singapore, January 23–March 16, 2020. *Morbidity and Mortality Weekly Report.* 2020;69:411–415. doi:10.15585/mmwr.mm6914e1.

36. Tong Z-D, Tang A, Li K-F, et al. Potential presymptomatic transmission of SARS-CoV-2, Zhejiang Province, China, 2020. *Emerging Infectious Diseases.* 2020;26(5):1052–1054. doi:10.3201/eid2605.200198.

37. There have been rare reports of asymptomatic SARS patients. Che X-Y, Di B, Zhao G-P, et al. A patient with asymptomatic severe acute respiratory syndrome (SARS) and antigenemia from the 2003–2004 community outbreak of SARS in Guangzhou, China. *Clinical Infectious Diseases.* 2006;43(1):e1–e5. doi:10.1086/504943.

38. Greenhalgh T, Jimenez JL, Prather KA, et al. Ten scientific reasons in support of airborne transmission of SARS-CoV-2. *Lancet.* 2021;397(10285):1603–1605. doi:10.1016/S0140-6736(21)00869-2.

39. Cepelewicz J. Chasing the elusive numbers that define epidemics. *Quanta Magazine.* March 22, 2021. https://www.quantamagazine.org/chasing-covid-19s-r0-and-other-numbers-that-define-epidemics-20210322/#. Accessed July 15, 2021.

40. Li Q, Guan X, Wu P, et al. Early transmission dynamics in Wuhan, China, of novel coronavirus–infected pneumonia. *New England Journal of Medicine.* 2020;382(13): 1199–1207. doi:10.1056/NEJMoa2001316.

41. Chan J F-W, Yuan S, Kok K-H, et al. A familial cluster of pneumonia associated with the 2019 novel coronavirus indicating person-to-person transmission: a study of a family cluster. *Lancet.* 2020;395(10223):514–523. doi:10.1016/S0140-6736(20)30154-9.

42. Sanche S, Lin Y T, Xu C, et al. High contagiousness and rapid spread of severe acute respiratory syndrome coronavirus 2. *Emerging Infectious Diseases.* 2020;26(7): 1470–1476. doi:10.3201/eid2607.200282.

43. Xia F, Yang X, Cheke RA. Quantifying competitive advantages of mutant strains in a population involving importation and mass vaccination rollout. *Infectious Disease Modelling.* 2021;6:988–996. doi:10.1016/j.idm.2021.08.001.

44. Van Kerkhove M. @mvankerkhove COVID: WHO renames UK and other variants with Greek letters. BBC News. May 31, 2021. https://www.bbc.com/news/world-57308592. Accessed August 26, 2021.

45. Brown CM, Vostok J, Johnson H, et al. Outbreak of SARS-CoV-2 infections, including COVID-19 vaccine breakthrough infections, associated with large public gatherings—Barnstable County, Massachusetts, July 2021. *Morbidity and Mortality Weekly Report.* 2021;70(31):1059–1062. doi:10.15585/mmwr.mm7031e2.

46. Doucleff M. The Delta variant isn't as contagious as chickenpox. But it's still highly contagious. National Public Radio. August 11, 2021. https://www.npr.org/sections/goatsandsoda/2021/08/11/1026190062/covid-delta-variant-transmission-cdc-chickenpox. Accessed August 26, 2021.

47. Liu Y, Rocklov J. The effective reproductive number of the Omicron variant of SARS-CoV-2 is several times relative to Delta. *Journal of Travel Medicine.* 2022;29(3):-taac037. doi:10.1093/jtm/taac037.

48. Macola IG. Timeline: how COVID-19 spread aboard the *Diamond Princess* cruise ship. Ship Technology. August 4, 2020. https://www.ship-technology.com/features/timeline-covid-spread-aboard-diamond-princess-cruise-ship/. Accessed August 1, 2021.

49. Sekizuka T, Itokawa K, Kageyama T, et al. Haplotype networks of SARS-CoV-2 infections in the *Diamond Princess* cruise ship outbreak. *Proceedings of the National Academy of Science.* 2020;117(33):20198–20201. doi:10.1073/pnas.2006824117.

50. Nishiura H. Backcalculating the incidence of infection with COVID-19 on the *Diamond Princess. Journal of Clinical Medicine.* 2020;9(3):657. doi:10.3390/jcm9030657.

51. Sekizuka T, Itokawa K, Kageyama T, et al. Haplotype networks of SARS-CoV-2 infections in the *Diamond Princess* cruise ship outbreak. *Proceedings of the National Academy of Science.* 2020;117(33):20198–20201. doi:10.1073/pnas.2006824117.

52. Moriarty LF, Plucinski MM, Marston BJ, et al. Public Health Responses to COVID-19 Outbreaks on Cruise Ships—Worldwide, February–March 2020. *Morbidity and Mortality Weekly Report.* 2020;69:347–352. doi:10.15585/mmwr.mm6912e3.

53. Mallapaty S. What the cruise-ship outbreaks reveal about COVID-19. *Nature.* March 26, 2020. https://www.nature.com/articles/d41586-020-00885-w. Accessed August 1, 2021.

54. Liu Y, Eggo RM, Kucharski AJ. Secondary attack rate and superspreading events for SARS-CoV-2. *Lancet.* 2020;395(10227):e47. doi:10.1016/S0140-6736(20)30462-1.

55. Lewis D. Superspreading drives the COVID pandemic—and could help to tame it. *Nature.* February 23, 2021. https://www.nature.com/articles/d41586-021-00460-x#ref-CR1. Accessed August 2, 2021.

56. Zimmer C. One meeting in Boston seeded tens of thousands of infections, study finds. *New York Times.* August 26, 2020. https://www.nytimes.com/2020/08/26/health/covid-19-superspreaders-boston.html. Accessed August 3, 2021.

57. Lemieux JE, Siddle KJ, Shaw BM et al. Phylogenetic analysis of SARS-CoV-2 in Boston highlights impact of superspreading events. *Nature.* 2021;371(6529):eabe3261. doi:10.1126/science.abe3261.

58. Xiao F, Tang M, Zheng X, et al. Evidence for gastrointestinal infection of SARS-CoV-2. *Gastroenterology*. 2020;158(6):1831–1833.e3. doi:10.1053/j.gastro.2020.02.055.

59. Xu Y, Li X, Zhu B, et al. Characteristics of pediatric SARS-CoV-2 infection and potential evidence for persistent fecal viral shedding. *Nature Medicine*. 2020;26:502–505. doi:10.1038/s41591-020-0817-4.

60. Hellmer M, Paxeus N, Magnius L, et al. Detection of pathogenic viruses in sewage provided early warnings of hepatitis A virus and norovirus outbreaks. *Applied and Environmental Microbiology*. 2014;80(21):6771–6781. doi:10.1128/AEM.01981-14.

61. Poyry T, Stenvik M, Hovi T. Viruses in sewage waters during and after a poliomyelitis outbreak and subsequent nationwide oral poliovirus vaccination campaign in Finland. *Applied and Environmental Microbiology*. 1988;54(2):371–374. doi:10.1128/aem.54.2.371–374.1988.

62. Medema G, Heijnen L, Elsinga G, et al. Presence of SARS-Coronavirus-2 RNA in sewage and correlation with reported COVID-19 prevalence in the early stage of the epidemic in the Netherlands. *Environmental Science and Technology Letters*. 2020;7:511–516. doi:10.1021/acs.estlett.0c00357.

63. Mallapaty S. How sewage could reveal true scale of coronavirus outbreak. *Nature*. 2020;580:176–177. doi:10.1038/d41586-020-00973-x.

64. Kreier F. The myriad ways sewage surveillance is helping fight COVID around the world. *Nature*. May 10, 2021. doi:10.1038/d41586-021-01234-1.

65. Harris-Lovett S, Nelson KL, Beamer P, et al. Wastewater surveillance for SARS-CoV-2 on college campuses: initial efforts, lessons learned, and research needs. *International Journal of Environmental Research and Public Health*. 2021;18(9):4455. doi:10.3390/ijerph18094455.

66. Albastali A, Naji M, Lootah R, et al. First confirmed detection of SARS-CoV-2 in untreated municipal and aircraft wastewater in Dubai, UAE: the use of wastewater-based epidemiology as an early warning tool to monitor the prevalence of COVID-19. *Science of the Total Environment*. 2021;760:143350. doi:10.1016/j.scitotenv.2020.143350.

67. Kreier F. The myriad ways sewage surveillance is helping fight COVID around the world. *Nature*. May 10, 2021. doi:10.1038/d41586-021-01234-1.

68. Van Doorn AS, Meijer B, Frampton CMA, et al. Systematic review with meta-analysis: SARS-CoV-2 stool testing and the potential for faecal-oral transmission. *Alimentary Pharmacology and Therapeutics*. 2020;52:1276–1288. doi:10.1111/apt.16036.

69. Cheung KS, Hung I, Chan P, et al. Gastrointestinal manifestations of SARS-CoV-2 infection and virus load in fecal samples from a Hong Kong cohort: systematic review and meta-analysis. *Gastroenterology*. 2020;159(1):81–95. doi:10.1053/j.gastro.2020.03.065.

70. Mohan N, Deswal S. Corona virus disease (COVID-19) fecal-oral transmission: is it a potential risk for Indians? *Indian Journal of Gastroenterology*. 2020;39(3):305–306. doi:10.1007/s12664-020-01072-5.

71. Dancer SJ, Li Y, Hart A, et al. What is the risk of acquiring SARS-CoV-2 from the use of public toilets? *Science of the Total Environment*. 2021;792:148341. doi:10.1016/j.scitotenv.2021.148341.

72. Service RF. Does disinfecting surfaces really prevent the spread of coronavirus? *Science*. March 12, 2020. doi:10.1126/science.abb7058.

73. WHO. Report of the WHO-China joint mission on coronavirus disease 2019 (COVID-19). February 16–24, 2020: 9. https://www.who.int/docs/default-source/coronaviruse/who-china-joint-mission-on-covid-19-final-report.pdf. Accessed August 11, 2021.

74. Van Doremalen N, Bushmaker T, Morris DH, et al. Aerosol and surface stability of SARS-CoV-2 as compared with SARS-CoV-1. *New England Journal of Medicine.* 2020;382(16):1564–1567. doi:10.1056/NEJMc2004973.

75. Lewis D. COVID-19 rarely spreads through surfaces. So why are we still deep cleaning? *Nature.* January 29, 2021. doi:10.1038/d41586-021-00251-4.

76. Goldman E. Exaggerated risk of transmission of COVID-19 by fomites. *Lancet Infectious Diseases.* 2020;20(8):892–893. doi:10.1016/S1473-3099(20)30561-2.

77. Marcenac P, Park GW, Duca LM, et al. Detection of SARS-CoV-2 on surfaces in households of persons with COVID-19. *International Journal of Environmental Research and Public Health.* 2021;18(15):8184. doi:10.3390/ijerph18158184.

78. WHO. Transmission of SARS-CoV-2: implications for infection prevention precautions. Scientific brief. July 9, 2020. https://www.who.int/news-room/commentaries/detail/transmission-of-sars-cov-2-implications-for-infection-prevention-precautions. Accessed August 12, 2021.

79. CDC. Science brief: SARS-CoV-2 and surface (fomite) transmission for indoor community environments. April 5, 2021. https://www.cdc.gov/coronavirus/2019-ncov/more/science-and-research/surface-transmission.html. Accessed August 12, 2021.

80. Tang JW, Marr LC, Li Y, et al. COVID-19 has redefined airborne transmission. *British Medical Journal.* 2021;373:n913. doi:10.1136/bmj.n913.

81. Jayaweera M, Perera H, Gunawardana B, et al. Transmission of COVID-19 virus by droplets and aerosols: a critical review on the unresolved dichotomy. *Environmental Research.* 2020;188:109819. doi:10.1016/j.envres.2020.109819.

82. Lee BU. Minimum sizes of respiratory particles capable of carrying SARS-CoV-2 and the possibility of aerosol generation. *International Journal of Environmental Research and Public Health.* 2020;17(19):6960. doi:10.3390/ijerph17196960.

83. Tang JW, Bahnfleth WP, Bluyssen PM, et al. Dismantling myths on the airborne transmission of severe acute respiratory syndrome coronavirus-2 (SARS-CoV-2). *Journal of Hospital Infection.* 2021;110:89–96. doi:10.1016/j.jhin.2020.12.022.

84. National Academies of Sciences. "Video 31-CQ1 reflection and syntheses: identifying opportunities and gaps on the path ahead by Kim Prather." Airborne Transmission of SARS-CoV-2: A Virtual Workshop, 26–27 August 2020. https://www.nationalacademies.org/event/08-26-2020/airborne-transmission-of-sars-cov-2-a-virtual-workshop. Accessed August 20, 2021.

85. Tang JW, Bahnfleth WP, Bluyssen PM, et al. Dismantling myths on the airborne transmission of severe acute respiratory syndrome coronavirus-2 (SARS-CoV-2). *Journal of Hospital Infection.* 2021;110:89–96. doi:10.1016/j.jhin.2020.12.022.

86. Tang JW, Bahnfleth WP, Bluyssen PM, et al. Dismantling myths on the airborne transmission of severe acute respiratory syndrome coronavirus-2 (SARS-CoV-2). *Journal of Hospital Infection.* 2021;110:89–96. doi:10.1016/j.jhin.2020.12.022.

87. Tang JW, Bahnfleth WP, Bluyssen PM, et al. Dismantling myths on the airborne transmission of severe acute respiratory syndrome coronavirus-2 (SARS-CoV-2). *Journal of Hospital Infection.* 2021;110:89–96. doi:10.1016/j.jhin.2020.12.022.

88. Prather KA, Marr LC, Schooley RT. Airborne transmission of SARS-CoV-2. *Science.* 2020;370(6514):303–304. doi:10.1126/science.abf0521.

89. Greenhalgh T, Jimenez JL, Prather KA, et al. Ten scientific reasons in support of airborne transmission of SARS-CoV-2. *Lancet*. 2021;397:1603–1605. doi:10.1016/S0140-6736(21)00869-2.

90. Tufekci Z. Why did it take so long to accept the facts about covid?" *New York Times*. May 7, 2021. https://www.nytimes.com/2021/05/07/opinion/coronavirus-airborne-transmission.html. Accessed August 22, 2021.

CHAPTER THREE: Humans

1. Kennedy JL, Turner RB, Braciale T, et al. Pathogenesis of rhinovirus infection. *Current Opinions in Virology*. 2012;2(3):287–293. doi:10.1016/j.coviro.2012.03.008; Kajon AE, Lu X, Erdman DD, et al. Molecular epidemiology and brief history of emerging adenovirus 14—associated respiratory disease in the United States. *Journal of Infectious Diseases*. 2010;202(1):93–103. doi:10.1086/653083; Hannoun C. The evolving history of influenza viruses and influenza vaccines. *Expert Review of Vaccines*. 2013;12(9):1085–1094. doi:10.1586/14760584.2013.824709; Henrickson KJ. Parainfluenza viruses. *Clinical Microbiology Reviews*. 2003;16(2):242–264. doi:10.1128/CMR.16.2.242-264.2003; Battles MB, McLellan JS. Respiratory syncytial virus entry and how to block it. *Nature Reviews Microbiology*. 2019;17:233–245. doi:10.1038/s41579-019-0149-x.

2. Richmond C. David Tyrrell. Obituaries. *British Medical Journal*. 2005;330(7505):1451. https://www.ncbi.nlm.nih.gov/pmc/articles/PMC558394/. Accessed September 30, 2021.

3. Tyrell DAJ. M.L. Bynoe. Obituary Notices. *British Medical Journal*. 1969;2:827. doi:10.1136/bmj.2.5660.827.

4. Kendall EJC. Letters: William Pickles lecture 1984. *Journal of the Royal College of General Practitioners*. 1984;Page 625. https://bjgp.org/content/bjgp/34/268/624.2.full.pdf. Accessed September 30, 2021.

5. Kendall EJC, Bynoe ML, Tyrrell DAJ. Virus Isolations From Common Colds Occurring in a Residential School. *British Medical Journal*. 1962;2(5297):82–86. http://www.jstor.org/stable/20374032.

6. Kronvall G, Nordenfelt E. Letter: on the history of human coronaviruses. *APMIS: Journal of Pathology, Microbiology and Immunology*. 2021;129(7):381–383. doi:10.1111/apm.13109.

7. Hoorn B. Respiratory viruses in model experiments. *Acta Oto-Laryngologica*. 1964;57(Suppl. 188):138–144. doi:10.3109/00016486409134552.

8. Tyrrell DAJ, Bynoe ML. Cultivation of a novel type of common-cold virus in organ cultures. *British Medical Journal*. 1965;1(5448):1467–1470. doi:10.1136/bmj.1.5448.1467.

9. Balaram P. "Discoverer of coronavirus." Frontline. May 22, 2020. https://frontline.thehindu.com/cover-story/article31510055.ece. Accessed September 29, 2021.

10. Hamre D, Procknow JJ. A new virus isolated from the human respiratory tract. *Proceedings of the Society for Experimental Biology and Medicine*. 1966;121:190–193. doi:10.3181/00379727-121-30734.

11. There is evidence that human coronavirus 229 E and porcine transmissible gastroenteritis virus are genetically related. Bridgen A, Duarte M, Tobler K, et al. Sequence determination of the nucleocapsid protein gene of the porcine epidemic diarrhoea virus confirms that this virus is a coronavirus related to human

coronavirus 229 E and porcine transmissible gastroenteritis virus. *Journal of General Virology.* 1993;74(9):1795–1804. doi:10.1099/0022-1317-74-9-1795.

12. Almeida JD, Tyrrell DAJ. The morphology of three previously uncharacterized human respiratory viruses that grow in organ culture. *Journal of General Virology.* 1967;1:175–178. doi:10.1099/0022-1317-1-2-175.

13. Almeida JD, Berry DM, Cunningham CH, et al. Virology: coronaviruses. *Nature.* 1968;220:650. doi:10.1038/220650b0.

14. Mahase E. COVID-19: first coronavirus was described in *The BMJ* in 1965. *British Medical Journal.* 2020;369:m1547. doi:10.1136/bmj.m1547.

15. Almeida JD, Berry DM, Cunningham CH, et al. Virology: coronaviruses. *Nature.* 1968;220:650. doi:10.1038/220650b0.

16. Gaunt ER, Hardie A, Class ECJ, et al. Epidemiology and clinical presentations of the four human coronaviruses 229 E, HKU1, NL 63, and OC 43 detected over 3 years using a novel multiplex real-time PCR method down-pointing small open triangle. *Journal of Clinical Microbiology.* 2010;48(8):2940–2947. doi:10.1128/JCM.00636-10.

17. Oransky I. Obituary. David Tyrrell. *Lancet.* 2005;365:2084. https://www.thelancet.com/pdfs/journals/lancet/PIIS0140673605667220.pdf. Accessed October 5, 2021.

18. Bradburne AF, Bynoe ML, Tyrrell DA. Effects of a "new" human respiratory virus in volunteers. *British Medical Journal.* 1967;3(5568):767–9. doi:10.1136/bmj.3.5568.767.

19. McIntosh K, Dees JH, Becker WB, et al. Recovery in tracheal organ cultures of novel viruses from patients with respiratory disease. *Proceedings of the National Academy of Science.* 1967;57(4):933–940. doi:10.1073/pnas.57.4.933.

20. McIntosh K, Becker WB, Chanock RM. Growth in suckling-mouse brain of "IBV-like" viruses from patients with upper respiratory tract disease. *Proceedings of the National Academy of Science.* 1967;58(6):2268–2273. doi:10.1073/pnas.58.6.2268.

21. Kay HS, Dowdle WR. Some characteristics of hemagglutination of certain strains of "IBV-Like" virus. *Journal of Infectious Diseases.* 1969;120(5):576–581. http://www.jstor.org/stable/30102206.

22. McIntosh, K. Personal interview via Zoom. October 11, 2021.

23. McIntosh K. Coronaviruses in the limelight. *Journal of Infectious Diseases.* 2005;191(4):489–491. doi:10.1086/428510.

24. Miller J. Website for the germ-obsessed. *Los Angeles Times.* January 13, 2007. https://www.latimes.com/news/la-oe-judymiller13jan13-story.html. Accessed October 18, 2021; Yu VL, Madoff LC. ProMED-mail: an early warning system for emerging diseases. *Clinical Infectious Diseases.* 2004;39(2):227–232. doi:10.1086/422003.

25. Cunnion, S. Senior fellow, medical director for the Center for Health Policy and Preparedness at Potomac Institute for Policy Studies, Arlington, VA. Personal communication via LinkedIn message. October 26, 2021.

26. ProMED. Pneumonia-China (Guangdong): RFI. ProMED-mail archive 20030210.0357. February 10, 2003. https://promedmail.org/promed-post/?id=2203342. Accessed October 18, 2021.

27. ProMED-mail. Pneumonia-China (Guangdong) Virus outbreak in southern China kills 5, prompts panic buying of antibiotics. ProMED-mail archive 20030211.0369. http://www.promedmail.org. Accessed October 18, 2021.

28. Xu RH, He JF, Evans MR, et al. Epidemiologic clues to SARS origin in China. *Emerging Infectious Diseases.* 2004;10(6):1030–1037. doi:10.3201/eid1006.030852.

29. Xu RH, He JF, Evans MR, et al. Epidemiologic clues to SARS origin in China. *Emerging Infectious Diseases*. 2004;10(6):1030–1037. doi:10.3201/eid1006.030852.

30. Engvall E, Perlmann P. Enzyme-linked immunosorbent assay, ELISA. *Journal of Immunology*. 1972;109(1):129–135. https://www.jimmunol.org/content/109/1/129. Accessed December 9, 2021.

31. CDC. Prevalence of IgG antibody to SARS-associated coronavirus in animal traders—Guangdong Province. *Morbidity and Mortality Weekly Report*. 2003;52(41):986–987. https://www.cdc.gov/mmwr/preview/mmwrhtml/mm5241a2.htm#tab1. Accessed December 9, 2021.

32. Kahn LH. Viral trade and global public health. *Issues in Science and Technology*. 2004;20(2). https://issues.org/kahn/. Accessed October 19, 2021.

33. Peiris JSM, Yuen KY, Osterhaus ADME, et al. The severe acute respiratory syndrome. *New England Journal of Medicine*. 2003;349:2431–41. doi:10.1056/NEJMra032498.

34. CDC. Preliminary clinical description of severe acute respiratory syndrome. March 28, 2003. 52(12):255–256. https://www.cdc.gov/mmwr/preview/mmwrhtml/mm5212a5.htm. Accessed October 19, 2021.

35. Peiris JSM, Yuen KY, Osterhaus ADME, et al. The severe acute respiratory syndrome. *New England Journal of Medicine*. 2003;349:2431–41. doi:10.1056/NEJMra032498.

36. *New England Journal of Medicine*. May 15, 2003;348(20). https://www.nejm.org/toc/nejm/348/20?query=article_issue_link. Accessed November 23, 2021.

37. Lee N, Hui D, Wu A, et al. A major outbreak of severe acute respiratory syndrome in Hong Kong. *New England Journal of Medicine*. 2003;348:1986–1994. doi:10.1056/NEJMoa030685.

38. Ksiazek TC, Erdman D, Goldsmith CS, et al. A novel coronavirus associated with severe acute respiratory syndrome. *New England Journal of Medicine*. 2003;348:1953–1966. doi:10.1056/NEJMoa030781; Drosten C, Gunther S, Preiser W, et al. Identification of a novel coronavirus in patients with severe acute respiratory syndrome. *New England Journal of Medicine*. 2003;348:1967–1976. doi:10.1056/NEJMoa030747.

39. National Human Genome Research Institute. Polymerase chain reaction (PCR) fact sheet. https://www.genome.gov/about-genomics/fact-sheets/Polymerase-Chain-Reaction-Fact-Sheet. Accessed October 22, 2021.

40. Saidi RK, Scharf S, Faloona F, Mullis KB, et al. Enzymatic amplification of β-globin genomic sequences and restriction site analysis for diagnosis of sickle cell anemia. *Science*. 1985;230(4732):1350–1354. doi:10.1126/science.2999980.

41. Emery SL, Erdman DD, Bowen MD, et al. Real-time reverse transcription-polymerase chain reaction assay for SARS-associated coronavirus. *Emerging Infectious Diseases*. 2004;10(2):311–316. doi:10.3201/eid1002.030759.

42. Lau SKP, Woo PCY, Wong BHL, et al. Detection of severe acute respiratory syndrome (SARS) coronavirus nucleocapsid protein in SARS patients by enzyme-linked immunosorbent assay. *Journal of Clinical Microbiology*. 2004;42(7):2884–2889. doi:10.1128/JCM.42.7.2884-2889.2004.

43. Cassedy A, Parle-McDermott A, O'Kennedy R. Virus detection: a review of the current and emerging molecular and immunological methods. *Frontiers in Molecular Biosciences*. 2021;8:637559. doi:10.3389/fmolb.2021.637559. The initial step of the ELISA test involved coating a diagnostic surface with the animal-derived antibodies. The second step added the clinical specimen (e.g., saliva, urine, or fecal

matter) to the diagnostic surface. The third step involved adding the second antibodies to the two other layers like adding the top part of a sandwich. When the two antibodies bound to their viral targets, they emitted a signal that could be detected. The ELISA test had drawbacks, however, namely cross-reactivity with other viruses, potentially resulting in false positive or negative results.

44. Fiers W, Contreras R, Duerinck F, et al. Complete nucleotide sequence of bacteriophage MS2 RNA: primary and secondary structure of the replicase gene. *Nature*. 1976;260(5551):500–507. doi:10.1038/260500a0.

45. Dugan VG, Saira K, Ghedin E. Large-scale sequencing and the natural history of model human RNA viruses. *Future Virology*. 2012;7(6):563–573. doi:10.2217/fvl.12.45.

46. He J-F, Peng G-W, Min J, et al. Molecular evolution of the SARS coronavirus during the course of the SARS epidemic in China. *Science*. 2004;303(5664):1666–1669. doi:10.1126/science.1092002.

47. Guan Y, Zheng BJ, Xe YQ, et al. Isolation and characterization of viruses related to the SARS coronavirus from animals in southern China. *Science*. 2003;302(5643):276–278. doi:10.1126/science.1087139.

48. He J-F, Peng G-W, Min J, et al. Molecular evolution of the SARS coronavirus during the course of the SARS epidemic in China. *Science*. 2004;303(5664):1666–1669. doi:10.1126/science.1092002.

49. Chiu RWK, Chim SSC, Tong Y-k, et al. Tracing SARS-coronavirus variant with large genomic deletion. *Emerging Infectious Diseases*. 2005;11(1):168–170. doi:10.3201/eid1101.040544.

50. Lam MH-B, Wing Y-K, Wai-Man M, et al. Mental morbidities and chronic fatigue in severe acute respiratory syndrome survivors. *Archives of Internal Medicine*. 2009;169(22):2142–2147. doi:10.1001/archinternmed.2009.384; Zhang P, Li J, Liu H, et al. Long-term bone and lung consequences associated with hospital-acquired severe acute respiratory syndrome: a 15-year follow-up from a prospective cohort study. *Bone Research*. 2020;8:8. doi:10.1038/s41413-020-0084-5.

51. Chew SK. SARS: how a global epidemic was stopped. *Bulletin of the World Health Organization*. 2007;85(4):324. doi:10.2471/BLT.07.032763.

52. Heymann DL. "How SARS was contained." *New York Times*. March 14, 2013. https://www.nytimes.com/2013/03/15/opinion/global/how-sars-was-contained.html. Accessed December 10, 2021.

53. Bamford C. "The original SARS virus disappeared—here's why coronavirus won't do the same." *Conversation*. June 5, 2020. https://theconversation.com/the-original-sars-virus-disappeared-heres-why-coronavirus-wont-do-the-same-138177. Accessed December 10, 2021.

54. CDC. SARS response timeline. https://www.cdc.gov/about/history/sars/timeline.htm. Accessed December 15, 2021.

55. Fouchier RAM, Hartwig NG, Bestebroer TM, et al. A previously undescribed coronavirus associated with respiratory disease in humans. *Proceedings of the National Academy of Science*. 2004;101(16):6212–6216. doi:10.1073/pnas.0400762101.

56. Pyrc K, Jebbink MF, Berkhout B, et al. Detection of new viruses by VIDISCA. Virus discovery based on cDNA-amplified fragment length polymorphism. *Methods in Molecular Biology*. 2004;454:73–89. doi:10.1007/978-1-59745-181-9_7. VIDISCA is the acronym for **V**irus ***Dis***covery based on ***c***DNA-**A**FLP (amplified fragment length polymorphism).

57. Van der Hoek L, Pyrc K, Jebbink MF, et al. Identification of a new human coronavirus. *Nature Medicine.* 2004;10(4):368–373. doi:10.1038/nm1024.

58. Fielding BC. Human coronavirus NL63: a clinically important virus? *Future Microbiology.* 2011;6(2):153–159. doi:10.2217/fmb.10.166.

59. Esper F, Shapiro ED, Weibel C, et al. Association between a novel human coronavirus and Kawasaki disease. *Journal of Infectious Diseases.* 2005;191(4):499–502. doi:10.1086/428291.

60. Tao Y, Shi M, Chommanard C, et al. Surveillance of bat coronaviruses in Kenya identifies relatives of human coronaviruses NL63 and 229 E and their recombination history. *Journal of Virology.* 2017;91(5):e01953–16. doi:10.1128/JVI.01953-16.

61. Woo PC, Lau SK, Chu CM, et al. Characterization and complete genome sequence of a novel coronavirus, coronavirus HKU1, from patients with pneumonia. *Journal of Virology.* 2005;79(2):884–95. doi:10.1128/JVI.79.2.884-895.2005.

62. Goes LG, Durigon EL, Campos AA, et al. Coronavirus HKU1 in children, Brazil, 1995. *Emerging Infectious Diseases.* 2011;17(6):1147–1148. doi:10.3201/eid1706.101381.

63. Woo PCY, Lau SKP, Yip CCY, et al. More and more coronaviruses: human coronavirus HKU1. *Viruses.* 2009;1(1):57–71. doi:10.3390/v1010057.

64. Hussein I. The story of the first MERS patient. *Nature Middle East.* June 2, 2014. https://www.natureasia.com/en/nmiddleeast/article/10.1038/nmiddleeast.2014.134. Accessed January 7, 2022. For his efforts to identify the novel coronavirus and notify ProMED, Dr. Zaki was forced to leave Saudi Arabia and return to Egypt.

65. ProMED. Novel coronavirus—Saudi Arabia: human isolate. ProMED-mail 2012; September 20. Archive Number: 20120920.1302733. https://promedmail.org/promed-post/?id=1302733. Accessed December 28, 2021.

66. Zaki AM, van Boheemen S, Bestebroer TM, et al. Isolation of a novel coronavirus from a man with pneumonia in Saudi Arabia. *New England Journal of Medicine.* 2012;367:1814–1820. doi:10.1056/NEJMoa1211721.

67. Illumina. https://www.illumina.com/. Accessed December 30, 2021.

68. Thermo Fisher Scientific. About. https://corporate.thermofisher.com/us/en/index/about.html. Accessed December 30, 2021.

69. Zaki AM, van Boheemen S, Bestebroer TM, et al. Isolation of a novel coronavirus from a man with pneumonia in Saudi Arabia. *New England Journal of Medicine.* 2012;367:1814–1820. doi:10.1056/NEJMoa1211721.

70. De Groot RJ, Baker SC, Baric RS, et al. Middle East respiratory syndrome coronavirus (MERS-CoV) announcement of the coronavirus study group. *Journal of Virology.* 2013;87(14):7790–7792. doi:10.1128/JVI.01244-13.

71. Drosten C, Seilmaier M, Corman VM, et al. Clinical features and virological analysis of a case of Middle East respiratory syndrome coronavirus infection. *Lancet Infectious Diseases.* 2013;13(9):745–751. doi:10.1016/S1473-3099(13)70154–3.

72. Muller MA, Meyer B, Corman VM, et al. Presence of Middle East respiratory syndrome coronavirus antibodies in Saudi Arabia: a nationwide, cross-sectional, serological study. *Lancet Infectious Diseases.* 2015;15(5):559–564. doi:10.1016/S1473-3099(15)70090-3.

73. Cevik M, Tate M, Lloyd O, et al. SARS-CoV-2, SARS-CoV, and MERS-CoV viral load dynamics, duration of viral shedding, and infectiousness: a systematic review and meta-analysis. *Lancet Microbe.* 2021;2(1):e13–e22. doi:10.1016/S2666–5247.

(20)30172-5; Memish ZA, Assiri AM, Al-Tawfiq JA. Middle East respiratory syndrome coronavirus (MERS-CoV) viral shedding in the respiratory tract: an observational analysis with infection control implications. *International Journal of Infectious Diseases*. 2014;29:307–308. doi:10.1016/j.ijid.2014.10.002.

74. Omrani AS, Shalhoub S. Middle East respiratory syndrome coronavirus (MERS-CoV): what lessons can we learn? *Journal of Hospital Infection*. 2015;91:188–196. doi:10.1016/j.jhin.2015.08.002.

75. World Health Organization. Regional Office for the Eastern Mediterranean. MERS situation update: January 2021. https://apps.who.int/iris/handle/10665/349420. Accessed January 4, 2022.

76. Ahmed H, Patel K, Greenwood DC, et al. Long-term clinical outcomes in survivors of severe acute respiratory syndrome and Middle East respiratory syndrome coronavirus outbreaks after hospitalization or ICU admission: a systematic review and meta-analysis. *Journal of Rehabilitation Medicine*. 2020;52(5):jrm00063. doi:10.2340/16501977-2694.

77. Lednicky JA, Tagliamonte MS, White SK, et al. Independent infections of porcine deltacoronavirus among Haitian children. *Nature*. 2021;600:133–137. doi:10.1038/s41586-021-04111-z.

78. Bose P, Mincer J, Buerkle T. The doctor whose gut instinct beat AI in spotting the coronavirus. Oliver Wyman Forum. March 5, 2020. https://www.oliverwymanforum.com/city-readiness/2020/mar/the-doctor-whose-gut-instinct-beat-ai-in-spotting-the-coronavirus.html. Accessed January 6, 2022.

79. ProMED. Undiagnosed pneumonia—China (Hubei): request for information. December 30, 2019. Archive Number: 20191230.6864153. https://promedmail.org/promed-post/?id=6864153. Accessed January 5, 2022.

80. WHO. COVID-19-China. January 5, 2020. https://www.who.int/emergencies/disease-outbreak-news/item/2020-DON229. Accessed January 7, 2022.

81. Huang C, Wang Y, Li X, et al. Clinical features of patients infected with 2019 novel coronavirus in Wuhan, China. *Lancet*. 2020;395(10223): P497–506. doi:10.1016/S0140-6736(20)30183–5.

82. Chen N, Zhou M, Dong X, et al. Epidemiological and clinical characteristics of 99 cases of 2019 novel coronavirus pneumonia in Wuhan, China: a descriptive study. *Lancet*. 2020;395:507–13. doi:10.1016/S0140–6736(20)30211-7.

83. Pekar J, Worobey M, Moshiri N, et al. Timing the SARS-CoV-2 index case in Hubei province. *Science*. 2021;372(6540):412–417. doi:10.1126/science.abf8003.

84. Ma J. Coronavirus: China's first confirmed COVID-19 case traced back to November 17. *South China Morning Post*. March 13, 2020. https://www.scmp.com/news/china/society/article/3074991/coronavirus-chinas-first-confirmed-covid-19-case-traced-back. Accessed January 21, 2022.

85. Liu A, Li Y, Wan Z. Seropositive prevalence of antibodies against SARS-CoV-2 in Wuhan, China. *JAMA Network Open*. 2020;3(10):e2025717. doi:10.1001/jamanetworkopen.2020.25717.

86. Li Z, Guan X, Mao N, et al. Antibody seroprevalence in the epicenter Wuhan, Hubei, and six selected provinces after containment of the first epidemic wave of COVID-19 in China. *Lancet Regional Health—Western Pacific*. 2021;8:100094. doi:10.1016/j.lanwpc.2021.100094.

87. Duan S, Zhou M, Zhang W, et al. Seroprevalence and asymptomatic carrier status of SARS-CoV-2 in Wuhan City and other places of China. *PLOS Neglected Tropical Diseases*. 2021;15(1):e0008975. doi:10.1371/journal.pntd.0008975.

88. Chang L, Hou W, Zhao L, et al. The prevalence of antibodies to SARS-CoV-2 among blood donors in China. *Nature Communication*. 2021;12. doi:10.1038/s41467-021-21503-x.

89. Pan Y, Li X, Yang G, et al. Seroprevalence of SARS-CoV-2 immunoglobulin antibodies in Wuhan, China: part of the city-wide massive testing campaign. *Clinical Microbiology and Infection*. 2021;27(2):253–257. doi:10.1016/j.cmi.2020.09.044.

90. He Z, Ren L, Yang J, et al. Seroprevalence and humoral immune durability of anti-SARS-Cov-2 antibodies in Wuhan, China: a longitudinal, population-level, cross-sectional study. *Lancet*. 2021;397(10279):1075–1084. doi:10.1016/S0140-6736(21)00238-5.

91. Worobey M. Dissecting the early COVID-19 cases in Wuhan. *Science*. 2021;374(6572):1202–1204. doi:10.1126/science.abm4454.

92. Zhang X, Tan Y, Ling Y, et al. Viral and host factors related to the clinical outcome of COVID-19. *Nature*. 2020;583:437–440. doi:10.1038/s41586-020-2355-0.

93. Wadman M, Couzin-Frankel J, Kaiser J, et al. A rampage through the body. *Science*. 2020;368(6489):356–360. doi:10.1126/science.368.6489.356.

94. Domingo P, Mur I, Pomar V, et al. The four horsemen of a viral apocalypse: the pathogenesis of SARS-CoV-2 infection (COVID-19). *eBioMedicine*. 2020;58:102887. doi:10.1016/j.ebiom.2020.102887.

95. Gupta A, Madhavan MV, Sehgal K, et al. Extrapulmonary manifestations of COVID-19. *Nature Medicine*. 2020;26:1017–1032. doi:10.1038/s41591-020-0968-3.

96. Shelton JF, Shastri AJ, Fletez-Brant K, et al. The UGT2A1/UGT2A2 locus is associated with COVID-19-related loss of smell or taste. *Nature Genetics*. 2022;54:121–124. doi:10.1038/s41588-021-00986-w.

97. Garduno-Soto M, Choreno-Parra JA, Cazarin-Barrientos J. Dermatological aspects of SARS-CoV-2 infection: mechanisms and manifestations. *Archives of Dermatological Research*. 2021;313(8):611–622. doi:10.1007/s00403-020-02156-0; Rongioletti F. SARS-CoV, MERS-CoV and COVID-19: what differences from a dermatological viewpoint? *Journal of the European Academy of Dermatology and Venereology*. 2020;34(10):e581–e582. doi:10.1111/jdv.16738.

98. Tenforde MW, Kim SS, Lindsell CJ, et al. Symptom duration and risk factors for delayed return to usual health among outpatients with COVID-19 in a multistate health care systems network—United States, March–June 2020. *Morbidity and Mortality Weekly Report*. 2020;69(30):993–998. doi:10.15585/mmwr.mm6930e1.

99. Taquet M, Dercon Q, Luciano S, et a. Incidence, co-occurrence, and evolution of long-COVID features: a 6-month retrospective cohort study of 273,618 survivors of COVID-19. *PLOS Medicine*. 2021;18(9):e1003773. doi:10.1371/journal.pmed.1003773.

100. Lopez-Leon S, Wegman-Ostrosky T, Perelman C, et al. More than 50 long-term effects of COVID-19: a systematic review and meta-analysis. *Scientific Reports*. 2021;11:16144. doi:10.1038/s41598-021-95565-8.

101. Van Kessel S, Olde Hartman TC, Lucassen P, et al. Post-acute and long COVID-19 symptoms in patients with mild diseases: a systematic review. *Family Practice*. 2022;39(1):159–167. doi:10.1093/fampra/cmab076.

102. Couzin-Frankel J. Clues to long COVID. *Science*. 2022;376(6599):1261–1265. https://www.science.org/doi/epdf/10.1126/science.add4297. Accessed August 17, 2022.

103. National Institutes of Health News Releases. NIH builds large nationwide study population of tens of thousands to support research on long-term effects of COVID-19. September 15, 2021. https://www.nih.gov/news-events/news-releases/nih

-builds-large-nationwide-study-population-tens-thousands-support-research-long -term-effects-covid-19. Accessed August 17, 2022.

CHAPTER FOUR: Molecular Biology of Coronaviruses

1. Messenger RNA has an inherent "handedness"(also called "chirality" or "polarity") to its structure. In general, living organisms have right (+) handedness. Class V viruses have negative (–) sense RNA that must be converted into positive (+) sense RNA to be used by the cell's machinery (i.e., ribosomes) to make new viruses.

2. Nobel Prize. David Baltimore: Facts. https://www.nobelprize.org/prizes /medicine/1975/baltimore/facts/. Accessed March 24, 2022; American Association of Immunologists. David Baltimore, PhD. https://www.aai.org/About/History/Notable -Members/Nobel-Laureates/DavidBaltimore. Accessed March 24, 2022. In 1975, Dr. David Baltimore was awarded the Nobel Prize in Physiology or Medicine for his co-discovery of retrovirus reverse transcriptase and the interactions between tumor viruses and cellular genetic material. In 1990, he became president of Rockefeller University, and in 1997, he became president of the California Institute of Technology.

3. Baltimore D. Expression of animal virus genomes. *Bacteriological Reviews*. 1971;35(3):235–241. doi:10.1128/br.35.3.235-241.1971.

4. Galibert F, Mandart E, Fitoussi F, et al. Nucleotide sequence of the hepatitis B virus genome (subtype ayw) cloned in E. coli. *Nature*. 1979;281:646–650. doi:10.1038 /281646a0.

5. In the Tree of Life, one hypothesis proposes that eukaryotic cells resulted from the merging of archaea and bacteria, which subsequently led to multicellular life. Spand A, Mahendrarahah TA, Offre P, et al. Evolving perspective on the origin and diversification of cellular life and the virosphere. *Genome Biology and Evolution*. 2022;14(6):evac034. doi:10.1093/gbe/evac034.

6. Woese CR and Fox GE. Phylogenetic structure of the prokaryotic domain: the primary kingdoms. *Proceedings of the National Academy of Science*. 1977;74(11):5088–5090. doi:10.1073/pnas.74.11.5088.

7. Koonin EV, Krupovic M, Agoi VI. The Baltimore classification of viruses 50 years later: how does it stand in the light of virus evolution? *Microbiology and Molecular Biology Reviews*. 2021;85(3):e0005321. doi:10.1128/MMBR.00053-21.

8. Molecules and organic chemical compounds such as DNA, RNA, and proteins typically have chirality, meaning that they have configurations that cannot be superimposed the same way that left and right hands cannot be superimposed on each other. They are mirror images. Similarly, positive- and negative-strand RNA might have the same properties, but they are chiral. Messenger RNA has positive chirality.

9. Koonin EV, Krupovic M, Agoi VI. The Baltimore classification of viruses 50 years later: how does it stand in the light of virus evolution? *Microbiology and Molecular Biology Reviews*. 2021;85(3):e0005321. doi:10.1128/MMBR.00053-21.

10. Stern A, Andino R. Chapter 17. Viral evolution: it is all about mutations. *Viral Pathogenesis (Third Edition). From Basics to Systems Biology*. Edited by Katze MG, Korth MJ, Law GL, Nathanson N. Academic Press, 2016. Pages 233–240. doi:10.1016/ B978-0-12-800964-2.00017-3.

11. Duffy S. Why are RNA virus mutation rates so damn high? *PLOS Biology*. 2018;16(8):e3000003. doi:10.1371/journal.pbio.3000003.

12. Sanjuan R, Nebot MR, Chirico N, et al. Viral mutation rates. *Journal of Virology*. 2010;84(19):9733–9748. doi:10.1128/JVI.00694-10.

13. Domingo E, Holland JJ. RNA virus mutations and fitness for survival. *Annual Review of Microbiology*. 1997;51:151–178. doi:10.1146/annurev.micro.51.1.151.

14. Woo PCY, Huang Y, Lau SKP, et al. Coronavirus genomics and bioinformatics analysis. *Viruses*. 2010;2(8):1804–1820. doi:10.3390/v2081803.

15. Lauber C, Goeman JJ, del Carmen Parquet M, et al. The footprint of genome architecture in the largest genome expansion in RNA viruses. *PLOS Pathogens*. 2013;9(7):e1003500. doi:10.1371/journal.ppat.1003500.

16. Obermeyer F, Jankowiak M, Barkas N, et al. Analysis of 6.4 million SARS-CoV-2 genomes identifies mutations associated with fitness. *Science*. 2022. doi:10.1126/science.abm1208.

17. Arakawa H. Mutation signature of SARS-CoV-2 variants raises questions to their natural origins. Zenodo. 2022. doi:10.5281/zenodo.6601991.

18. Carstens EB. Ratification vote on taxonomic proposals to the International Committee on Taxonomy of Viruses (2009). *Archives of Virology*. 2010;155(1):133–146. doi:10.1007/s00705-009-0547-x.

19. Forni D, Cagliani R, Clerici M, et al. Molecular evolution of human coronavirus genomes. *Trends in Microbiology*. 2017;25(1):35–48. doi:10.1016/j.tim.2016.09.001.

20. Benton MJ, Donoghue PCJ. Paleontological evidence to date the tree of life. *Molecular Biology and Evolution*. 2006;24(1):26–53. doi:10.1093/molbev/msl150.

21. Cui J, Li F, Shi Z-L. Origin and evolution of pathogenic coronaviruses. *Nature Reviews Microbiology*. 2019;17(3):181–192. doi:10.1038/s41579-018-0118-9.

22. Genomes such as DNA and RNA are read in one direction, analogous to English sentences being read from left to right. In the case of genomes, one end is written as the 5′ (5 prime) end. The opposite end is written as the 3′ (3 prime) end. Genomes are always read (translated) from the 5′ to the 3′ ends. By convention, genomes are written with the 5′ end to the left and the 3′ end to the right. Open reading frames (ORFs) are the parts of the genome that specify the amino acid sequence used to make a protein. Three nucleotides translate into a specific amino acid. Like the beginning of a sentence, the ORFs typically begin with an ATG and end with termination codes such as TAG, TGA, or TAA. Brown TA. Chapter 7. Understanding a genome sequence. *Genomes*. 2nd ed. Oxford: Wiley-Liss; 2002. https://www.ncbi.nlm.nih.gov/books/NBK21136/.

23. V'kovski PV, Kratzel A, Steiner S, et al. Coronavirus biology and replication: implications for SARS-CoV-2. *Nature Reviews Microbiology*. 2021;19:155–170. doi:10.1038/s41579-020-00468-6.

24. Kaur N, Singh R, Dar Z, et al. Genetic comparison among various coronavirus strains for the identification of potential vaccine targets of SARS-CoV-2. *Infection, Genetics and Evolution*. 2021;89:104490. doi:10.1016/j.meegid.2020.104490.

25. White JM, Delos SE, Brecher M, et al. Structures and mechanisms of viral membrane fusion proteins. *Critical Reviews in Biochemistry and Molecular Biology*. 2009;43(3):189–219. doi:10.1080/10409230802058320.

26. Harrison SC. Viral membrane fusion. *Virology*. 2015;479–480:498–507. doi:10.1016/j.virol.2015.03.043.

27. Russell CJ, Hu M, Okda FA. Influenza hemagglutinin protein stability, activation, and pandemic risk. *Trends in Microbiology*. 2018;26(10):841–853. doi:10.1016/j.tim.2018.03.005.

28. Spike proteins have been divided into three classes based on their structure and activation processes. Class I viruses, including influenza, HIV-1, and

coronaviruses, typically have spike proteins composed of three parts, called a trimer, derived from a single-chain precursor. These spike proteins require cleavage by host cell enzymes called proteases to be activated for the fusion process. Class II viruses such as alphaviruses, bunyaviruses, and flaviviruses require the cleavage of a second viral surface protein that acts as a "chaperone" for the first fusion protein. Class III viruses, including rhabdoviruses and herpesviruses, combine features of the first two classes. In some cases, there is no obvious priming event and the process can be reversible. For more information: Harrison SC. Viral membrane fusion. *Nature Structural & Molecular Biology*. 2008;15:690–698. doi:10.1038/nsmb.1456.

29. Gary R. Whittaker, PhD, professor of virology, Department of Microbiology and Immunology, Cornell University College of Veterinary Medicine. Email communication May 2, 2022.

30. Salamanna F, Maglio M, Landini MP, et al. Body localization of ACE-2: on the trail of the keyhole of SARS-CoV-2. *Frontiers in Medicine*. 2020;7:594495. doi:10.3389/fmed.2020.594495.

31. Damas J, Hughes GM, Keough KC, et al. Broad host range of SARS-CoV-2 predicted by comparative and structural analysis of ACE2 in vertebrates. *Proceedings of the National Academy of Science*. 2020;117(36):22311–22322. doi:10.1073/pnas.2010146117.

32. Du L, He Y, Zhou Y, et al. The spike protein of SARS-CoV—a target for vaccine and therapeutic development. *Nature Reviews Microbiology*. 2009;7:226–236. doi:10.1038/nrmicro2090.

33. Du L, He Y, Zhou Y, et al. The spike protein of SARS-CoV—a target for vaccine and therapeutic development. *Nature Reviews Microbiology*. 2009;7:226–236. doi:10.1038/nrmicro2090.

34. Matsuyama S, Ujike M, Morikawa S, et al. Protease-mediated enhancement of severe acute respiratory syndrome coronavirus infection. *Proceedings of the National Academy of Science*. 2005;102(35):122543–12547. doi:10.1073/pnas.0503203102.

35. Thomas G. Furin at the cutting edge: from protein traffic to embryogenesis and disease. *Nature Review Molecular Cell Biology*. 2002;3(10):753–766.

36. He Z, Khatib A-M, Creemers JWM. The proprotein convertase furin in cancer: more than an oncogene. *Oncogene*. 2022;41:1252–1262. doi:10.1038/s41388-021-02175–9.

37. Kim SH, Wang R, Gordon DJ, et al. Furin mediates enhanced production of fibrillogenic Abri peptides in familial British dementia. *Nature Neuroscience*. 1999;2(11):984–988. doi:10.1038/14783.

38. Klenk H-D, Garten W. Host cell proteases controlling virus pathogenicity. *Trends in Microbiology*. 1994;2(2):39–43. doi:10.1016/0966-842X(94)90123–6.

39. Wu Y, Zhao S. Furin cleavage sites naturally occur in coronaviruses. *Stem Cell Research*. 2021;50:102115. doi:10.1016/j.scr.2020.102115.

40. Klenk H-D, Garten W. Host cell proteases controlling virus pathogenicity. *Trends in Microbiology*. 1994;2(2):39–43. doi:10.1016/0966-842X(94)90123–6.

41. Vey M, Orlich M, Adler S, et al. Hemagglutinin activation of pathogenic avian influenza viruses of serotype H7 requires the protease recognition motif R-X-K/R-R. *Virology*. 1992;188:408–413. doi:10.1016/0042-6822(92)90775-K.

42. Cavanagh D, Davis PJ, Pappin DJ, et al. Coronavirus IBV: partial amino terminal sequencing of spike polypeptide S2 identifies the sequence Arg-Arg-Phe-Arg-

Arg at the cleavage site of the spike precursor propolypeptide of IBV strains Beaudette and M41. *Virus Research.* 1986;4(2):133–143. doi:10.1016/0168-1702(86)90037-7.

43. Luczo JM, Tachedjian M, Harper JA, et al. Evolution of high pathogenicity of H5 avian influenza virus: haemagglutinin cleavage site selection of reverse-genetics mutants during passage in chickens. *Scientific Reports.* 2018;8:11518. doi:10.1038/s41598-018-29944-z.

44. Bergeron E, Vincent MJ, Wickham L, et al. Implication of proprotein convertases in the processing and spread of severe acute respiratory syndrome coronavirus. *Biochemical and Biophysical Research Communications.* 2005;326(3):554–563. doi:10.1016/j.bbrc.2004.11.063.

45. Follis KE, York J, Nunberg JH. Furin cleavage of the SARS coronavirus spike glycoprotein enhances cell-cell fusion but does not affect virion entry. *Virology.* 2006;350(2):358–369. doi:10.1016/j.virol.2006.02.003.

46. Watanabe R, Matsuyama S, Shirato K, et al. Entry from the cell surface of severe acute respiratory syndrome coronavirus with cleaved S protein as revealed by pseudotype virus bearing cleaved S protein. *Journal of Virology.* 2008;82(23):11985–11991. doi:10.1128/JVI.01412-08.

47. Belouzard S, Chu VC, Whittaker GR. Activation of the SARS coronavirus spike protein via sequential proteolytic cleavage at two distinct sites. *Proceedings of the National Academy of Science.* 2009;106(14):5871–5876. doi:10.1073/pnas.0809524106.

48. Johnson BA, Xie X, Bailey AL, et al. Loss of furin cleavage site attenuates SARS-CoV-2 pathogenesis. *Nature.* 2021;591:293–299. doi:10.1038/s41586-021-03237-4.

49. Raj V, Mou H, Smits S, et al. Dipeptidyl peptidase 4 is a functional receptor for the emerging human-EMC. *Nature.* 2013;495:251–254. doi:10.1038/nature12005.

50. Wang N, Shi X, Jiang L, et al. Structure of MERS-CoV spike receptor-binding domain complexed with human receptor DPP4. *Cell Research.* 2013;23(8):986–993. doi:10.1038/cr.2013.92.

51. Meyerholz DK, Lambertz AM, McCray, Jr. PB. Dipeptidyl peptidase 4 distribution in the human respiratory tract. *American Journal of Pathology.* 2016;186(1):78–86. doi:10.1016/j.ajpath.2015.09.014.

52. Deacon CF. Physiology and pharmacology of DPP-4 in glucose homeostasis and the treatment of type 2 diabetes. *Frontiers in Endocrinology.* 2019;10:80. doi:10.3389/fendo.2019.00080.

53. Millet JK, Whittaker GR. Host cell entry of Middle East respiratory syndrome coronavirus after two-step, furin-mediated activation of the spike protein. *Proceedings of the National Academy of Science.* 2014;111(42):15214–15219. doi:10.1073/pnas.1407087111.

54. Beyerstedt S, Casaro EB, Rangel EB. COVID-19: angiotensin-converting enzyme 2 (ACE2) expression and tissue susceptibility to SARS-CoV-2 infection. *European Journal of Clinical Microbiology and Infectious Diseases.* 2020;40:905–919. doi:10.1007/s10096-020-04138-6.

55. Samavati L, Uhal BD. ACE2, much more than just a receptor for SARS-CoV-2. *Frontiers in Cellular and Infection Microbiology.* 2020;10. doi:10.3389/fcimb.2020.00317.
ACE2 is the acronym for angiotensin-converting enzyme 2, which is a membrane-bound glycoprotein consisting of 805 amino acids. In contrast to ACE2, angiotensin-converting enzyme (ACE) is a protein that regulates the renin-angiotensin-aldosterone system. ACE2 and ACE share 42 percent of their amino acids but have different structures and functions in the body.

56. Bestle D, Heindl MR, Limburg H, et al. TMPRSS2 and furin are both essential for proteolytic activation of SARS-CoV-2 in human airway cells. *Life Science Alliance.* 2020;3(9):e202000786. doi:10.26508/lsa.202000786.

57. Ou X, Liu Y, Lei X, et al. Characterization of spike glycoprotein of SARS-CoV-2 on virus entry and its immune cross-reactivity with SARS-CoV. *Nature Communication.* 2020;11:1620. doi:10.1038/s41467-020-15562-9.

58. Guruprasad L. Human coronavirus spike protein-host receptor recognition. *Progress in Biophysics and Molecular Biology.* 2021;161:39–53. doi:10.1016/j.pbiomolbio.2020.10.006.

59. Jackson CB, Farzan M, Chen B, et al. Mechanisms of SARS-CoV-2 entry into cells. *Nature Reviews Molecular Cell Biology.* 2022;23(1):3–20. doi:10.1038/s41580-021-00418-x.

60. Ord M, Faustova I, Loog M. The sequence at spike S1/S2 site enables cleavage by furin and phosphor-regulation in SARS-CoV2 but not in SARS-CoV1 or MERS-CoV. *Nature Scientific Reports.* 2020;10:16944. doi:10.1038/s41598-020-74101-0.

61. Segreto R, Deigin Y. The genetic structure of SARS-CoV-2 does not rule out a laboratory origin. *Bioessays.* 2020. doi:10.1002/bies.202000240.

62. Chan YA, Zhan SH. The emergence of the spike furin cleavage site in SARS-CoV-2. *Molecular Biology and Evolution.* 2021;39(1):msab327. doi:10.1093/molbev/msab327.

63. Dance A. The shifting sands of "gain-of-function" research. *Nature.* 2021;598:554–557. doi:10.1038/d41586-021-02903-x.

64. Fauci, Walensky COVID-19 response testimony senate hearing transcript July 20. Rev. July 20, 2021. https://www.rev.com/blog/transcripts/fauci-walensky-covid-19-response-testimony-senate-hearing-transcript-july-20. Accessed: November 10, 2021.

CHAPTER FIVE: Gain-of-Function Research, Biosafety, Biosecurity, and Bioethics

1. National Research Council. *Biotechnology research in an age of terrorism.* Washington, DC: National Academies Press; 2004:25–28.

2. Jackson RJ, Ramsay AJ, Christensen CD, et al. Expression of mouse interleukin-4 by a recombinant ectromelia virus suppresses cytolytic lymphocyte responses and overcomes genetic resistance to mousepox. *Journal of Virology.* 2001;75:1205–1210. doi:10.1128/JVI.75.3.1205-1210.2001.

3. Cello J, Paul AV, Wimmer E. Chemical synthesis of poliovirus cDNA: generation of infectious virus in the absence of natural template. *Science.* 2002;297(5583):1016–1018. doi:10.1126/science.1072266.

4. National Research Council. *Biotechnology research in an age of terrorism.* Washington, DC: National Academies Press; 2004:114–115.

5. Specter M. The trouble with scientific secrets. *New Yorker.* December 22, 2011. https://www.newyorker.com/news/news-desk/the-trouble-with-scientific-secrets. Accessed June 3, 2022.

6. Herfst S, Schrauwen EJA, Linster M, et al. Airborne transmission of influenza A/H5N1 virus between ferrets. *Science.* 2012;336(6088):1534–1541. doi:10.1126/science.1213362.

7. Fauci AS, Nabel GJ, Collins FS. A flu virus risk worth taking. *Washington Post.* December 30, 2011. https://www.washingtonpost.com/opinions/a-flu-virus-risk-worth-taking/2011/12/30/gIQAM9sNRP_story.html. Accessed June 3, 2022.

8. White House Office of Science and Technology Policy. A framework for guiding U.S. Department of Health and Human Services funding decisions about research proposals with the potential for generating highly pathogenic avian influenza H5N1 viruses that are transmissible among mammals by respiratory droplets. February 21, 2013. https://www.phe.gov/s3/dualuse/Documents/funding-hpai-h5n1.pdf. Accessed June 22, 2022.

9. Jackson DA, Symons RH, Berg P. Genetic information into DNA of simian virus 40: circular SV40 DNA molecules containing lambda phage genes and the galactose operon of *Escherichia coli*. *Proceedings of the National Academy of Sciences*. 1972;69(10):2904–2909. doi:10.1073/pnas.69.10.2904.

10. Berg P. Asilomar 1975: DNA modification secured. *Nature*. 2008;455:290–291. doi:10.1038/455290a.

11. Berg P, Singer MF. The recombinant DNA controversy: twenty years later. *Proceedings of the National Academy of Science*. 1995;92:9011–9013. doi:10.1073/pnas.92.20.9011.

12. Berg P. Potential biohazards of recombinant DNA molecules. *Proceedings of the National Academy of Science*. 1974;71(7):2593–2594. doi:10.1073/pnas.71.7.2593.

13. Berg P. Asilomar 1975: DNA modification secured. *Nature*. 2008;455:290–291. doi:10.1038/455290a.

14. Berg P. Asilomar 1975: DNA modification secured. *Nature*. 2008;455:290–291. doi:10.1038/455290a.

15. US Department of Health, Education, and Welfare. Recombinant DNA research guidelines. National Institutes of Health. *Federal Register*. 1976;41(131):27902–27943. https://osp.od.nih.gov/wp-content/uploads/2014/12/NIH_Guidelines_1976_Rev0.pdf. Accessed June 7, 2022.

16. Barbeito MS, Kruse RH. A history of the American Biological Safety Association part I: the first 10 Biological Safety Conferences 1955–1965. American Biological Safety Association. https://absa.org/about/hist01/. Accessed June 7, 2022.

17. Connell N. Biological agents in the laboratory—the regulatory issues. Public Interest Report. Federation of American Scientists. Fall 2011. https://pubs.fas.org/pir/2011fall/2011fall-bioagents.pdf. Accessed June 7, 2022.

18. Smart JK. History of chemical and biological warfare: an American perspective. In Sidell FR, Takafuji ET, Franz DR, eds. *Textbook of military medicine*. Washington DC: Office of the Surgeon General, Borden Institute; 1997:59–64.

19. Nixon's actions prompted the international community to develop a supplement to the 1925 Geneva Protocol, known as the Biological Weapons Convention (BWC). On April 10, 1972, the BWC was opened for signature. It was the first multilateral disarmament treaty that banned the production and use of an entire category of weapons. It prohibited the acquisition, development, production, transfer, stockpiling, and most importantly, use of biological and toxin weapons. (UN Office for Disarmament Affairs. Biological Weapons Convention. https://www.un.org/disarmament/biological-weapons/. Accessed June 8, 2022.) Unfortunately, the treaty lacked teeth. It had no formal mechanism to monitor compliance, which led to distrust between nations, as evidenced by the Soviet Union, which did not believe that the United States had terminated its offensive program. The Soviets built their own enormous offensive program. In 1989, one of the USSR's top scientists, a microbiologist named Vladimir Pasechnik, defected to the West, and three years later, another Soviet defector, Kanatjan Alibekov (Ken Alibek) secretly immigrated to the United States.

20. Kruse RH, Barbeito MS. A history of the American Biological Safety Association. part III: safety conferences 1978–1987. ABSA International. https://absa.org/about/hist03/. Accessed June 8, 2022.

21. Kruse RH, Barbeito MS. A history of the American Biological Safety Association. part III: safety conferences 1978–1987. ABSA International. https://absa.org/about/hist03/. Accessed June 8, 2022.

22. Centers for Disease Control and Prevention. *Biosafety in microbiological and biomedical laboratories (BMBL).* 6th ed. https://www.cdc.gov/labs/BMBL.html. Accessed June 8, 2022.

23. Byers KB, Harding AL. Chapter 4. Laboratory-associated infections. In: Fleming DO, Hunt DL, eds. *Biological safety: principles and practices.* 5th ed. Washington DC: ASM Press;2017:pp. 62–63, tables 2 and 3, p. 70.

24. Byers KB, Harding AL. Chapter 4. Laboratory-associated infections. In: Fleming DO, Hunt DL, eds. *Biological safety: principles and practices.* 5th ed. Washington DC: ASM Press;2017:pp. 70–74, table 5.

25. US Department of Health and Human Services. Science Safety Security. Biosecurity. https://www.phe.gov/s3/BioriskManagement/biosecurity/Pages/default.aspx. Accessed June 8, 2022.

26. Torok TJ, Taauxe RV, Wise RP, et al. A large community outbreak of salmonellosis caused by intentional contamination of restaurant salad bars. *JAMA.* 1997;278(5):389–395. https://www.cdc.gov/phlp/docs/forensic_epidemiology/Additional%20Materials/Articles/Torok%20et%20al.pdf. Accessed June 10, 2022; Federal Bureau of Investigation. Amerithrax or anthrax investigation. https://www.fbi.gov/history/famous-cases/amerithrax-or-anthrax-investigation. Accessed June 10, 2022.

27. CDC and USDA Federal Select Agent Program. 2020 annual report of the Federal Select Agent Program. https://www.selectagents.gov/. Accessed June 8, 2022.

28. US Department of Health and Human Services. Public Health Emergency. Science Safety Security. Biosecurity history. https://www.phe.gov/s3/BioriskManagement/biosecurity/Pages/History.aspx. Accessed June 8, 2022.

29. CDC and USDA Federal Select Agent Program. History. https://www.selectagents.gov/overview/history.htm. Accessed June 8, 2022.

30. CDC and USDA Federal Select Agent Program. Select agents and toxins list. https://www.selectagents.gov/sat/list.htm. Accessed June 8, 2022.

31. CDC and USDA Federal Select Agent Program. 2015 annual report. June 2016:5; 2016 annual report. October 2017:27; 2017 annual report. December 2018:24; 2018 annual report. January 2020:23; 2019 annual report. September 2020:24; 2020 annual report. September 2021:23. https://www.selectagents.gov/resources/publications/annualreport/2020.htm. Accessed June 8, 2022.

32. CDC and USDA Federal Select Agent Program. 2020 annual report. September 2021:11, 12, 23, 24. https://www.selectagents.gov/resources/publications/annualreport/2020.htm. Accessed June 8, 2022.

33. CDC and USDA Federal Select Agent Program. 2020 annual report. September 2021:8, 15, 16.

34. CDC and USDA Federal Select Agent Program. 2020 annual report. September 2021:16 and 17.

35. US Government Accountability Office. High-containment laboratories: coordinated actions needed to enhance the Select Agent Program's oversight of

hazardous pathogens. October 2017. GAO-18-145. https://www.gao.gov/assets/gao-18-145.pdf. Accessed June 9, 2022.

36. The comparable countries were Canada, France, Germany, the Netherlands, Switzerland, and the United Kingdom.

37. US Government Accountability Office. High-containment laboratories: coordinated actions needed to enhance the Select Agent Program's oversight of hazardous pathogens. October 2017. GAO-18-145. https://www.gao.gov/assets/gao-18-145.pdf. Accessed June 9, 2022.

38. Xia H, Yuan Z. High-containment facilities and the role they play in global health security. *Journal of Biosafety and Biosecurity*. 2021;4:1–4. doi:10.1016/j.jobb.2021.11.005.

39. Lentzos F, Koblentz GD. Mapping maximum biological containment labs globally. Maximum biocontainment policy brief. King's College London. May 2021. https://www.globalbiolabs.org/policy-brief. Accessed June 10, 2022.

40. Lentzos F, Koblentz GD. Mapping maximum biological containment labs globally. Maximum biocontainment policy brief. King's College London. May 2021. https://www.globalbiolabs.org/policy-brief. Accessed June 10, 2022.

41. Jacobsen R. Inside the risky bat-virus engineering links America to Wuhan. *MIT Technology Review*. June 29, 2021. https://www.technologyreview.com/2021/06/29/1027290/gain-of-function-risky-bat-virus-engineering-links-america-to-wuhan/. Accessed June 10, 2022.

42. Cohen J. Wuhan coronavirus hunter Shi Zhengli speaks out. *Science*. 2020;369(6503):487–488. doi:10.1126/science.369.6503.487; Shi Zhengli Q&A: Reply to Science magazine. *Science*. https://www.science.org/pb-assets/PDF/News%20PDFs/Shi%20Zhengli%20Q&A-1630433861.pdf. Accessed June 10, 2022.

43. Rogin J. State Department cables warned of safety issues at Wuhan lab studying bat coronaviruses. *Washington Post*. April 14, 2020. https://www.washingtonpost.com/opinions/2020/04/14/state-department-cables-warned-safety-issues-wuhan-lab-studying-bat-coronaviruses/. Accessed June 11, 2020.

44. Rogin J. State Department cables warned of safety issues at Wuhan lab studying bat coronaviruses. April 14, 2020. https://www.washingtonpost.com/opinions/2020/04/14/state-department-cables-warned-safety-issues-wuhan-lab-studying-bat-coronaviruses/. Accessed June 23, 2022.

45. US Department of Health and Human Services. National biodefense strategy. https://www.phe.gov/Preparedness/biodefense-strategy/Pages/default.aspx. Accessed June 12, 2022.

46. Cao C. China's evolving biosafety/biosecurity legislations. *Journal of Law and the Biosciences*. 2021;8(1):1–21. doi:10.1093/jlb/lsab020.

47. United Nations Office for Disarmament Affairs. Implementation Support Unit. https://www.un.org/disarmament/biological-weapons/implementation-support-unit. Accessed June 12, 2022.

48. World Health Organization. What we do. https://www.who.int/about/what-we-do. Accessed June 12, 2022.

49. Gostin LO, Katz R. The International Health Regulations: the governing framework for global health security. *Milbank Quarterly*. 2016;94(2):264–313. doi:10.1111/1468-0009.12186.

50. World Health Organization. *International Health Regulations*. 3rd ed. Geneva, Switzerland: WHO Press; 2005. https://www.who.int/publications/i/item/9789241580496, Accessed June 12, 2022.

51. World Health Organization. *International Health Regulations*. 3rd ed. Geneva, Switzerland: WHO Press; 2005. https://www.who.int/publications/i/item/9789241580496, Accessed June 12, 2022.

52. United Nations Office for Disarmament Affairs. UN Security Council Resolution 1540 (2004). https://www.un.org/disarmament/wmd/sc1540/. Accessed June 12, 2022.

53. Global Health Security Agenda. About the GHSA. https://ghsagenda.org/about-the-ghsa/. Accessed June 12, 2022.

54. Global Health Security Agenda. Member commitments. https://ghsagenda.org/member-commitments/. Accessed June 12, 2022.

55. Merriam-Webster Dictionary. Ethic (n.). https://www.merriam-webster.com/dictionary/ethics. Accessed June 14, 2022; Bioethics (n.). https://www.merriam-webster.com/dictionary/bioethics. Accessed June 14, 2022.

56. Harvard Law School Library. Nuremberg trials project. https://nuremberg.law.harvard.edu/. Accessed June 14, 2022.

57. Shuster E. Fifty years later: the significance of the Nuremberg Code. *New England Journal of Medicine*. 1997;337:1436–1440. doi:10.1056/NEJM199711133372006.

58. World Medical Association. WMA declaration of Helsinki—ethical principles for medical research involving human subjects. July 9, 2018. https://www.wma.net/policies-post/wma-declaration-of-helsinki-ethical-principles-for-medical-research-involving-human-subjects/. Accessed June 14, 2022.

59. CDC. The U.S. public health service syphilis study at Tuskegee: research implications. https://www.cdc.gov/tuskegee/after.htm. Accessed June 14, 2022.

60. Health and Human Services. Office for Human Research Protections. The Belmont report: ethical principles and guidelines for the protection of human subjects of research. https://www.hhs.gov/ohrp/regulations-and-policy/belmont-report/index.html. Accessed June 14, 2022.

61. National Archives. Code of Federal Regulations. Title 45 Public welfare. Subtitle A Department of Health and Human Services. Subchapter A General Administration. Part 46 Protection of Human Subjects. https://www.ecfr.gov/current/title-45/subtitle-A/subchapter-A/part-46?toc=1. Accessed June 14, 2022.

62. Grady C. Institutional Review Boards: purpose and challenges. *Chest*. 2015;148(5):1148–1155. doi:10.1378/chest.15-0706.

63. National Institutes of Health. Office of Laboratory Animal Welfare. The Institutional Animal Care and Use Committee (IACUC). https://olaw.nih.gov/resources/tutorial/iacuc.htm) Accessed June 14, 2022.

64. National Institutes of Health, NIH Director. Statement on funding pause on certain types of gain-of-function research. October 16, 2014. https://www.nih.gov/about-nih/who-we-are/nih-director/statements/statement-funding-pause-certain-types-gain-function-research. Accessed June 16, 2022; National Institutes of Health. US government gain-of-function deliberative process and research funding pause on selected gain-of-function research involving influenza, MERS, and SARS viruses. October 17, 2014. https://www.phe.gov/s3/dualuse/Documents/gain-of-function.pdf. Accessed June 22, 2022.

65. Kaiser J, Malakoff D. U.S. halts funding for new risky virus studies, calls for voluntary moratorium. *Science*. October 17, 2014. https://www.science.org/content/article/us-halts-funding-new-risky-virus-studies-calls-voluntary-moratorium. Accessed June 16, 2022.

66. National Institutes of Health. A report of the National Science Advisory Board for Biosecurity: recommendations for the evaluation and oversight of proposed gain-of-function research. May 2016:1–2, 15, 16. https://osp.od.nih.gov/wp-content/uploads/2016/06/NSABB_Final_Report_Recommendations_Evaluation_Oversight_Proposed_Gain_of_Function_Research.pdf. Accessed June 17, 2022.

67. Department of Health and Human Services. Recommended policy guidance for departmental development of review mechanisms for potential pandemic pathogen care and oversight (P3CO). January 9, 2017. https://www.phe.gov/s3/dualuse/Documents/P3CO-FinalGuidanceStatement.pdf. Accessed June 17, 2022. Some of these guidelines were derived from a previous 2012 government document, *United States Government Policy for the Oversight of Life Sciences Dual Use Research of Concern*. https://www.phe.gov/s3/dualuse/documents/us-policy-durc-032812.pdf. Accessed June 22, 2022.

68. National Institutes of Health. NIH lifts funding pause on gain-of-function research. December 19, 2017. https://www.nih.gov/about-nih/who-we-are/nih-director/statements/nih-lifts-funding-pause-gain-function-research. Accessed June 22, 2022.

69. National Institutes of Health. News and events: research involving enhanced potential pandemic pathogens. February 28, 2022. https://www.nih.gov/news-events/research-involving-potential-pandemic-pathogens. Accessed January 19, 2023.

70. National Institutes of Health. RePORTER. https://reporter.nih.gov/. Accessed January 19, 2023.

71. National Institutes of Health RePORTER. Understanding the risk of bat coronavirus emergence. Project leader: Peter Daszak. Project Numbers: 2R01AI1110964-06 and 5R01AI110964-05. Awardee organization: EcoHealth Alliance, Inc. https://reporter.nih.gov/search/xQW6UJmWfUuOV01ntGvLwQ/project-details/9819304 and https://reporter.nih.gov/search/xQW6UJmWfUuOV01ntGvLwQ/project-details/9491676. Accessed June 22, 2022.

72. Menachery VD, Yount BL Jr, Debbink K, et al. A SARS-like cluster of circulating bat coronaviruses shows potential for human emergence. *Nature Medicine*. 2015;21:1508–1513. doi:10.1038/nm.2985.

73. Wade N. The origin of COVID: did people or nature open Pandora's box at Wuhan? *Bulletin of the Atomic Scientists*. May 5, 2021. https://thebulletin.org/2021/05/the-origin-of-covid-did-people-or-nature-open-pandoras-box-at-wuhan/. Accessed June 22, 2022.

74. Butler D. Engineered bat virus stirs debate over risky research. *Nature*. 2015. doi:10.1038/nature.2015.18787.

75. Segreto R, Deigin Y. The genetic structure of SARS-CoV-2 does not rule out a laboratory origin. *BioEssays*. 2021;43:2000240. doi:10.1002/bies.202000240.

76. Fernandez A. Molecular biology clues portray SARS-CoV-2 as a gain-of-function laboratory manipulation of RaTG13. *ACS Medical Chemistry Letters*. 2021;12(6):941–942. doi:10.1021/acsmedchemlett.1c00274.

77. Racaniello V. This week in virology 615: Peter Daszak of EcoHealth Alliance. Interview December 19, 2019. YouTube. Posted May 19, 2020. 27:53 to 30:20. https://www.youtube.com/watch?v=IdYDL_RK--w. Accessed June 22, 2022.

78. Subbaraman N. "Heinous!": coronavirus researcher shut down for Wuhan-lab link slams new funding restrictions." *Nature*. August 21, 2020. https://www.nature.com/articles/d41586-020-02473-4. Accessed June 23, 2022.

79. Editorial Board. One person who might know what really happened in Wuhan. October 25, 2021. https://www.washingtonpost.com/opinions/2021/10/25/one-person-who-might-know-what-really-happened-wuhan/. Accessed June 23, 2022.

80. Goldmacher S. Republicans capture control of the House after falling short of midterm expectations." *New York Times*. Updated December 5, 2022. https://www.nytimes.com/live/2022/11/16/us/election-news-results. Accessed January 19, 2023.

81. Romm T. House GOP to embark on sweeping probe of COVID origin, U.S. response. *Washington Post*. January 9, 2023. https://www.washingtonpost.com/us-policy/2023/01/09/house-covid-origin-investigation/. Accessed January 19, 2023.

82. Kaiser J. House approves ban on gain-of-function pathogen research. *Science*. November 15, 2023. https://www.science.org/content/article/house-approves-ban-gain-function-pathogen-research. Accessed November 19, 2023.

83. Braun R. The public's growing distrust of science? *Nature Biotechnology*. 1999;17(Supplement 5):14. doi:10.1038/70359.

84. Achenbach J. Lab-leak fears are putting virologists under scrutiny. *Washington Post*. January 18, 2023. https://www.washingtonpost.com/health/2023/01/18/lab-leak-theory-virologist/. Accessed January 20, 2023.

CHAPTER SIX: Conclusions and Policy Recommendations

1. Kahn LH. Coronavirus or antibiotic resistance: our appetite for animals (wild and domestic) poses big disease risks. *Bulletin of the Atomic Scientists*. February 14, 2020 https://thebulletin.org/2020/02/think-chinas-wet-markets-for-wildlife-spread-diseases-industrial-meat-production-is-worse/#post-heading. Accessed August 25, 2022.

2. Ritchie H, Roser M. Meat and dairy production. Our World in Data. August 2017. https://ourworldindata.org/meat-production. Accessed August 25, 2022.

3. Nungesser F, Winter M. Meat and social change. *Österreichische Zeitschrift für Soziologie*. 2021;46:109–124. doi:10.1007/s11614-021-00453-0.

4. McCarthy J, Dekoster S. Nearly one in four in U.S. have cut back on eating meat. Gallup. January 27, 2020. https://news.gallup.com/poll/282779/nearly-one-four-cut-back-eating-meat.aspx. Accessed August 25, 2022.

5. Yeo S. Can the psycological technique of "pre-conformity" help change our harmful behaviors? *Pacific Standard*. August 14, 2018. https://psmag.com/environment/peer-pressure-can-make-americans-eat-less-meat. Accessed August 25, 2022.

6. Kahn LH. Confronting zoonoses, linking human and veterinary medicine. *Emerging Infectious Diseases*. 2006;12(4):556–561. doi:10.3201/eid1204.050956.

7. Worldometer. World population. https://www.worldometers.info/. Accessed January 20, 2023.

8. United Nations, Department of Economic and Social Affairs. World population projected to reach 9.8 billion in 2050, and 11.2 billion in 2100. https://www.un.org/en/desa/world-population-projected-reach-98-billion-2050-and-112-billion-2100. Accessed January 20, 2023.

9. Kahn LH. Developing a one health approach by using a multi-dimensional matrix. *One Health*. 2021;12:100289. doi:10.1016/j.onehlt.2021.100289.

10. Kahn LH. Air quality surveillance and control: how to improve the response to all respiratory diseases, COVID-19 included. *Bulletin of the Atomic Scientists*. September 8, 2021. https://thebulletin.org/2021/09/air-quality-surveillance-and-control-how-to-improve-the-response-to-all-respiratory-diseases-covid-19-included/#post-heading. Accessed August 25, 2022.

11. Tang JW, Bahnfleth WP, Bluyssen PM, et al. Dismantling myths on the airborne transmission of severe acute respiratory syndrome coronavirus-2 (SARS-CoV-2). *Journal of Hospital Infection*. 2021;110:89–96. doi:10.1016/j.jhin.2020.12.022.

12. Morawska L, Allen J, Bahnfleth W, et al. A paradigm shift to combat indoor respiratory infection. *Science*. 2021;372(6543):689–691. doi:10.1126/science.abg2025.

13. Morawska L, Allen J, Bahnfleth W, et al. A paradigm shift to combat indoor respiratory infection. *Science*. 2021;372(6543):689–691. doi:10.1126/science.abg2025.

14. Kahn LH. Biodefense research: can secrecy and safety coexist? *Biosecurity and Bioterrorism*. 2004;2(2):81–5. doi:10.1089/153871304323146379.

15. Matthews D. New global body aims to improve biosecurity and biosafety. *Science Business*. January 17, 2023. https://sciencebusiness.net/news/new-global-body-aims-improve-biosecurity-and-biosafety. Accessed November 19, 2023.

16. National Research Council. *Biotechnology research in an age of terrorism*. Washington, DC: National Academies Press; 2004. doi:10.17226/10827.

17. Department of Health and Human Services. National Institutes of Health. National Institute of Allergy and Infectious Diseases. Grant Number 1R01AI110964-01. Principal Investigator: Peter Daszak, PhD. Project Title: Understanding the risk of bat coronavirus emergence. Project Period: 06/01/2014-05/31/2019. Pp. 14, 92, 114, 116–117, 124–126, 131–133, 145, 160–161, 195, 197, 210, 230, 253, 272–273, 274, 298–299, 331, 334, 337, 361, 363–364, 378, 397, 484. https://www.documentcloud.org/documents/21055989-understanding-risk-bat-coronavirus-emergence-grant-notice. Accessed August 31, 2022.

18. Department of Health and Human Services. National Institutes of Health. National Institute of Allergy and Infectious Diseases. Grant Number 1R01AI110964-01. Principal Investigator: Peter Daszak, PhD. Project Title: Understanding the risk of bat coronavirus emergence. Project Period: 06/01/2014-05/31/2019. Pp. 459.

19. Department of Health and Human Services. National Institutes of Health. National Institute of Allergy and Infectious Diseases. Grant Number 1R01AI110964-01. Principal Investigator: Peter Daszak, PhD. Project Title: Understanding the risk of bat coronavirus emergence. Project Period: 06/01/2014-05/31/2019. Pp. 298, 361, 364, 378, 402.

20. Callier V. Machine learning in evolutionary studies comes of age. *Proceedings of the National Academy of Science*. 2022;119(17):e2205058119. doi:10.1073/pnas.2205058119.

21. Pavlova A, Zhang Z, Acharya A, et al. Machine learning reveals the critical interactions for SARS-CoV-2 spike protein binding to ACE2. *Journal of Physical Chemistry Letters*. 2021;12(23):5494–5502. doi:10.1021/acs.jpclett.1c01494.

22. Baeshen NA, Baeshen MN, Sheikh A, et al. Cell factories for insulin production. *Microbial Cell Factories*. 2014;2(13):141. doi:10.1186/s12934-014-0141-0.

APPENDIX 4: Discovering Messenger RNA and the Messenger RNA Vaccines

1. Cobb M. Who discovered messenger RNA? *Current Biology*. 2015;25:R523–R548. doi:10.1016/j.cub.2015.05.032.

2. Brenner S, Jacob F, Meselson M. An unstable intermediate carrying information from genes to ribosomes for protein synthesis. *Nature*. 1961;190:576–581. doi:10.1038/190576a0.

3. Gros F, Hiatt H, Gilbert W, et al. Unstable ribonucleic acid revealed by pulse labelling of Escherichia coli. 1961;190:581–585. doi:10.1038/190581a0.

4. Karikó K, PhD, adjunct professor of neurosurgery, University of Pennsylvania, email communication, August 2, 2021.

5. Dimitriadis G. Translation of rabbit globin mRNA introduced by liposomes into mouse lympthocytes. *Nature.* 1978;274:923–924. doi:10.1038/274923a0.

6. Krieg PA, Melton DA. Functional messenger RNAs are produced by SP6 in vitro transcription of cloned cDNAs. *Nucleic Acids Research.* 1984;12(18):7057–7070. doi:10.1093/nar/12.18.7057.

7. Martinon F, Krishnan S, Lenzen G, et al. Induction of virus-specific cytotoxic T lymphocytes in vivo by liposome-entrapped mRNA. *European Journal of Immunology.* 1993;23(7):1719–1722. doi:10.1002/eji.1830230749.

8. Zhou X, Berglund P, Rhodes G, et al. Self-replicating Semliki Forest virus RNA as recombinant vaccine. *Vaccine.* 1994;12(16):1510–1514. doi:10.1016/0264-410x(94)90074-4.

9. Takeda K, Kaisho T, Akira S. Toll-like receptors. *Annual Review of Immunology.* 2003;21:335–376. doi:10.1146/annurev.immunol.21.120601.141126.

10. Kariko K, Buckstein M, Ni H, et al. Suppression of RNA recognition by toll-like receptors: the impact of nucleoside modification and the evolutionary origin of RNA. *Immunity.* 2005;23:165–175. doi:10.1016/j.immuni.2005.06.008.

11. Kariko K, Muramatsu H, Welsh FA, et al. Incorporation of pseudouridine into mRNA yields superior nonimmunogenic vector with increased translational capacity and biological stability. *Molecular Therapy.* 2008;16(11):1833–1840. doi:10.1038/mt.2008.200.

12. Nair P. QnAs with Katalin Karikó. *Proceedings of the National Academy of Science.* 2021;118(51):e2119757118. doi:10.1073/pnas.2119757118.

13. Malone RW, Felgner PL, Verma IM. Cationic liposome-mediated RNA transfection. *Proceedings of the National Academy of Science.* 1989;86(16):6077–6081. doi:10.1073/pnas.86.16.6077.

14. Dolgin E. The tangled history of mRNA vaccines. *Nature.* September 14, 2021. https://www.nature.com/articles/d41586-021-02483-w#ref-CR1. Accessed July 8, 2022.

15. Anderson EJ, Rouphael NG, Widge AT, et al. Safety and immunogenicity of SARS-CoV-2 mRNA-1273 vaccine in older adults. *New England Journal of Medicine.* 2020;383(25):2427–2438. doi:10.1056/NEJMoa2028436.

16. FDA News Release. FDA takes key action in fight against COVID-19 by issuing emergency use authorization for first COVID-19 vaccine. December 11, 2020. https://www.fda.gov/news-events/press-announcements/fda-takes-key-action-fight-against-covid-19-issuing-emergency-use-authorization-first-covid-19. Accessed July 10, 2022.

17. FDA News Release. FDA takes additional action in fight against COVID-19 by issuing emergency use authorization for second COVID-19 vaccine. December 18, 2020. https://www.fda.gov/news-events/press-announcements/fda-takes-additional-action-fight-against-covid-19-issuing-emergency-use-authorization-second-covid. Accessed July 10, 2022.

18. Shinomiya N, Minari J, Yoshizawa G, et al. Reconsidering the need for gain-of-function research on enhanced potential pandemic pathogens in the post-COVID-19 era. *Frontiers in Bioengineering and Biotechnology.* 2022;10:966586. doi:10.3389/fbioe.2022.966586.

References

Appendix 1b.

Alexander TJL, Richards WPC, Roe CK. An encephalomyelitis of suckling pigs in Ontario. *Canadian Journal of Comparative Medicine Veterinary Science.* 1959;23(10):316–319. https://www.ncbi.nlm.nih.gov/pmc/articles/PMC1582309/pdf/vetsci00107-0008.pdf. Accessed February 12, 2021.

Artika M, Desantari AK, Wiyatno A. Molecular biology of coronaviruses: current knowledge. *Heliyon.* 2020;6:e04743. doi:10.1016/j.heliyon.2020.e04743.

Bande F, Arshad SS, Bejo MH, et al. Progress and challenges toward the development of vaccines against avian infectious bronchitis. *Journal of Immunology Research.* 2015:424860. doi:10.1155/2015/424860.

Binn LN, Lazar EC, Keenan KP, et al. Recovery and characterization of a coronavirus from military dogs with diarrhea. *Proceedings, Annual Meeting of the United States Animal Health Association.* 1974;78:359–66.

Boileau MJ, Kapil S. Bovine coronavirus associated syndromes. *Veterinary Clinics of North America: Food Animal Practice.* 2010;26(1):123–146. doi:10.1016/j.cvfa.2009.10.003.

Bradburne AF, Bynoe ML, Tyrrell DAJ. Effects of a "new" human respiratory virus in volunteers. *British Medical Journal.* 1967;3:767–769. https://www.bmj.com/content/bmj/3/5568/767.full.pdf. Accessed February 12, 2021.

Buonavoglia C, Decaro N, Martella V, et al. Canine coronavirus highly pathogenic for dogs. *Emerging Infectious Diseases.* 2006;12(3):492–494. doi:10.3201/eid1203.050839.

Carstens EB. Ratification vote on taxonomic proposals to the International Committee on Taxonomy of Viruses (2009). *Archives of Virology.* 2010;155(1):133–146. doi:10.1007/s00705-009-0547-x.

Cheever FS, Daniels JB, Pappenheimer AM, et al. A murine virus (JHM) causing disseminated encephalomyelitis with extensive destruction of myelin. I. Isolation and biological properties of the virus. *Journal of Experimental Medicine.* 1949;90:181–194.

Clark MA. Bovine coronavirus. *British Veterinary Journal.* 1993;149(1):51–70. doi:10.1016/S0007-1935(05)80210-6.

Cook JKA, Jackwood M, Jones RC. The long view: 40 years of infectious bronchitis research. *Avian Pathology.* 2012;41(3):239–250. doi:10.1080/03079457.2012.680432.

Dong N, Fang L, Zeng S, et al. Porcine deltacoronavirus in mainland China. *Emerging Infectious Diseases.* 2015;21(12):2254–2255. doi:10.3201/eid2112.150283.

Doyle LP, Hutchings LM. A transmissible gastroenteritis in pigs. *Journal of the American Veterinary Medical Association.* 1946;108:257–259.

Ellis J. Review article: what is the evidence that bovine coronavirus is a biologically significant respiratory pathogen in cattle? *Canadian Veterinary Journal.* 2019;60:147–152. https://www.ncbi.nlm.nih.gov/pmc/articles/PMC6340311/. Accessed February 17, 2021.

Ennaji Y, Khataby K, Ennaji MM. Infectious bronchitis virus in poultry: molecular epidemiology and factors leading to the emergence and reemergence of novel strains of infectious bronchitis virus. *Emerging and Reemerging Viral Pathogens.* 2020:31–44. doi:10.1016/B978-0-12-814966-9.00003-2.

Furseth H. Coronavirus: an emerging threat? *Equus Magazine.* January 27, 2020. https://equusmagazine.com/diseases/coronavirus-emerging-threat-28983. Accessed March 20, 2021.

Goes LG, Durigon EL, Campos AA, et al. Coronavirus HKU1 in children, Brazil, 1995. *Emerging Infectious Diseases.* 1995;17(6):1147–1148. doi:10.3201/eid1706.101381.

Greig AS, Johnson M, Bouillant AMP. Encephalomyelitis of swine caused by a haemagglutinating virus. VI. Morphology of the virus. *Research in Veterinary Science.* 1971;12(4):305–309. doi:10.1016/S0034-5288(18)34153–5.

Greig AS, Mitchell D, Corner AH, et al. A hemagglutinating virus producing encephalomyelitis in baby pigs. *Canadian Journal of Comparative Medicine Veterinary Science.* 1962;26(3):49–56. https://www.ncbi.nlm.nih.gov/pmc/articles/PMC1583410/pdf/vetsci00064-0003.pdf. Accessed February 12, 2021.

Guy J. Coronaviral enteritis of turkeys. *Merck Veterinary Manual.* 2020. https://www.merckvetmanual.com/poultry/viral-enteritis/coronaviral-enteritis-of-turkeys. Accessed March 11, 2021.

Guy JS, Breslin JJ, Breuhaus B, et al. Characterization of a coronavirus isolated from a diarrheic foal. *Journal of Clinical Microbiology.* 2000;38(12):4523–4526.

Holzworth JE. Some important disorders of cats. *Cornell Veterinarian.* 1963;53:157–160. https://babel.hathitrust.org/cgi/pt?id=uc1.b3779838&view=1up&seq=170. Accessed February 12, 2021.

Jaiswal NK, Saxena SK. Classical coronaviruses. In: Saxena S., ed. *Coronavirus disease 2019 (COVID-19): epidemiology, pathogenesis, diagnosis, and therapeutics.* Singapore: Springer; 2020. doi:10.1007/978-981-15-4814-7_12.

Jung K, Saif LJ, Wang Q. Porcine epidemic diarrhea virus (PEDV): an update on etiology, transmission, pathogenesis, and prevention and control. *Virus Research.* 2020;286:198045. doi:10.1016/j.virusres.2020.198045; USDA. Technical note: porcine epidemic diarrhea (PED). https://www.aasv.org/aasv%20website/Resources/Diseases/PED/usda_ped_tech_note.pdf. Accessed February 16, 2021.

Laude H, van Reeth K, Pensaert M. Porcine respiratory coronavirus: molecular features and virus-host interactions. *Veterinary Research.* 1993;24(2):125–150.

Le Poder S. Feline and canine coronaviruses: common genetic and pathobiological features. *Advances in Virology.* 2011:609465. doi:10.1155/2011/609465.

Marthaler D, Jiang Y, Collins J, et al. Complete genome sequence of strain SDCV/USA/Illinois121/2014, a porcine deltacoronavirus from the United States. *Genome Announcements.* 2014;2(2):e00218–14. doi:10.1128/genomeA.00218-14.

Mebus CA, Stair EL, Rhodes MB, et al. Neonatal calf diarrhea: propagation, attenuation and characteristics of a coronavirus-like agent. *American Journal of Veterinary Research.* 1973;34:145–50.

Mebus CA, White RG, Stair EL, et al. Neonatal calf diarrhea: results of a field trial using a reo-like virus vaccine. *Veterinary Medicine, Small Animal Clinician.* 1972;67(2):173–174 passim.

Niederwerder MC, Hesse RA. Swine enteric coronavirus disease: a review of 4 years of porcine epidemic diarrhoea virus and porcine deltacoronavirus in the United states and Canada. *Transboundary Emerging Diseases*. 2018;65(3):660–675. doi:10.1111/tbed.12823.

O'Reilly KJ, Fishman B, Hitchcock LM. Feline infectious peritonitis: isolation of a coronavirus. *Veterinary Record*. 1979;104(15):348. doi:10.1136/vr.104.15.348.

Parker JC, Cross SS, Rowe WP. Rat coronavirus (RCV): a prevalent, naturally occurring pneumotropic virus of rats. *Archiv Für die gesamte Virusforschung*. 1970;31:293–302. doi:10.1007/BF01253764.

Pensaert MB, De Bouck P. A new coronavirus-like particle associated with diarrhea in swine. *Archives of Virology*. 1978;58:243–247. doi:10.1007/BF01317606.

Peterson EH, Hymas TA. Antibiotics in the treatment of an unfamiliar turkey disease. *Poultry Science*. 1951;30(3):466–468. doi:10.3382/ps.0300466.

Ramakrishnan S, Kappala D. Avian infectious bronchitis virus. *Recent Advances in Animal Virology*. 2019:301–319. doi:10.1007/978-981-13-9073-9_16.

Randolph JF. Transmissible gastroenteritis. *Iowa State University Veterinarian*. 1963;26(2):article 5. https://lib.dr.iastate.edu/iowastate_veterinarian/vol26/iss2/5.

Saif LJ. Bovine respiratory coronavirus. *Veterinary Clinics of North America: Food Animal Practice*. 2010;26(2):349–364. doi:10.1016/j.cvfa.2010.04.005. https://www.ncbi.nlm.nih.gov/pmc/articles/PMC4094360/.

Schalk AF, Hawn MC. An apparently new respiratory disease of baby chicks. *Proceedings: thirty-fourth annual meeting of the United States Live Stock Sanitary Association*. Chicago, Illinois. December 1930: pp. 413–423. https://www.usaha.org/upload/Proceedings/1930-1959/1930_THIRTY_FOURTH_ANNUAL_MEETIN.pdf. Accessed October 1, 2020.

Schalk AF. An apparently new respiratory disease of baby chicks. *Journal of the American Veterinary Medical Association*. 1931;78:413–423.

Schikora BM, Shih LM, Hietala SK. Genetic diversity of avian infectious bronchitis virus California variants isolated between 1988 and 2001 based on the S1 subunit of the spike glycoprotein. *Archives of Virology*. 2003;148:115–136. doi:10.1007/s00705-002-0904-5.

Stevenson GW, Hoang H, Schwartz KJ et al. Emergence of porcine epidemic diarrhea virus in the United States: clinical signs, lesions, and viral genomic sequences. *Journal of Veterinary Diagnostic Investigation*. 2013;25(5):649–654. doi:10.1177/1040638713501675.

Sun R-Q, Cai R-J, Chen Y-Q, et al. Outbreak of porcine epidemic diarrhea in suckling piglets, China. *Emerging Infectious Diseases*. 2012;18(1):161–163. doi:10.3201/eid1801.111259.

Turlewicz-Podbielska H, Pomorska-Mol M. Porcine coronaviruses: overview of the state of the art. *Virologica Sinica*. 2021;36:833–851. doi:10.1007/s12250-021-00364-0.

US Centers for Disease Control and Prevention. Animals and COVID-19. Updated April 27, 2022. https://www.cdc.gov/coronavirus/2019-ncov/daily-life-coping/animals.html. Accessed May 9, 2022.

US Department of Agriculture. Confirmed cases of SARS-CoV-2 in animals in the United States. Updated May 2, 2022. https://www.aphis.usda.gov/aphis/ourfocus/onehealth/one-health-sarscov2-in-animals. Accessed May 9, 2022.

Virology: coronaviruses. *Nature*. 1968;220:650. doi:10.1038/220650b0.

Wood EN. An apparently new syndrome of porcine epidemic diarrhea. *Veterinary Record*. 1977;100(12):243–4. doi:10.1136/vr.100.12.243.

Zhou P, Fan H, Ma J-Y, et al. Fatal swine acute diarrhoea syndrome caused by an HKU2-related coronavirus of bat origin. *Nature*. 2018;556:255–258. doi:10.1038/s41586-018-0010-9.

Appendix 2.

Burki TK. Omicron variant and booster COVID-19 vaccines. *Lancet*. 2022;10(2):e17. doi:10.1016/S2213-2600(21)00559–2.

Gomes da Silva P, Nascimento MSJ, Soares RRG, et al. Airborne spread of infectious SARS-CoV-2: moving forward using lessons from SARS-CoV and MERS-Cov. *Science of the Total Environment*. 2021;764:142802. doi:10.1016/j.scitotenv.2020.142802.

Ke R, Romero-Severson E, Sanche S, et al. Estimating the reproductive number R_0 of SARS-CoV-2 in the United States and eight European countries and implications for vaccination. *Journal of Theoretical Biology*. 2021;517:110621. doi:10.1016/j.jtbi.2021.110621.

Liu Y, Rocklov J. The effective reproductive number of the omicron variant of SARS-CoV-2 is several times relative to delta. *Journal of Travel Medicine*. 2022;29(3):taac037. doi:10.1093/jtm/taac037.

Liu Y, Rocklov J. The reproductive number of the delta variant of SARS-CoV-2 is far higher compared to the ancestral SARS-CoV-2 virus. *Journal of Travel Medicine*. 2021:1–3. doi:10.1093/jtm/taab124.

Appendix 3a

SARS

CDC. Severe acute respiratory syndrome (SARS) archived website. https://www.cdc.gov/sars/surveillance/absence.html. Accessed January 12, 2022.

Chan-Yeung M, Rui-Heng XU. SARS: epidemiology. *Respirology*. 2003;8(Suppl 1):S9–S14. doi:10.1046/j.1440-1843.2003.00518.x.

Chu KH, Tsang WK, Tang CS, et al. Acute renal impairment in coronavirus-associated severe acute respiratory syndrome. *Kidney International*. 2005;67(2):698–705. doi:10.1111/j.1523-1755.2005.67130.x.

Donnelly CA, Ghani AC, Leung GM, et al. Epidemiological determinants of spread of causal agent of severe acute respiratory syndrome in Hong Kong. *Lancet*. 2003;361(9371):1761–1766. doi:10.1016/S0140-6736(03)13410–1.

Hui D S-C, Wong P-C, Wang C. SARS: clinical features and diagnosis. *Respirology*. 2003;8:S20–S24. doi:10.1046/j.1440-1843.2003.00520.x.

Lam M H-B, Wing Y-K, Wai-Man M, et al. Mental morbidities and chronic fatigue in severe acute respiratory syndrome survivors. *Archives of Internal Medicine*. 2009;169(22):2142–2147. doi:10.1001/archinternmed.2009.384.

Lee N, Hui D, Wu A, et al. A major outbreak of severe acute respiratory syndrome in Hong Kong. *New England Journal of Medicine*. 2003;348:1986–94. doi:10.1056/NEJMoa030685.

Zhang P, Li J, Liu H, et al. Long-term bone and lung consequences associated with hospital-acquired severe acute respiratory syndrome: a 15-year follow-up from a prospective cohort study. *Bone Research*. 2020;8:8. doi:10.1038/s41413-020-0084-5.

MERS

Alhogbani T. Acute myocarditis associated with novel Middle East respiratory syndrome coronavirus. *Annals Saudi Medicine* 2016;36:78–80. doi:10.5144/0256-4947.2016.78.

Assiri A, Al-Tawfiq JA, Al-Rabeeah AA, et al. Epidemiological, demographic, and clinical characteristics of 47 cases of Middle East respiratory syndrome coronavirus disease from Saudi Arabia: a descriptive study. *Lancet Infectious Diseases* 2013;13(9):752–761. doi:10.1016/S1473-3099(13)70204–4.

CDC. MERS clinical features. https://www.cdc.gov/coronavirus/mers/clinical-features.html. Accessed January 11, 2022.

Omrani AS, Shalhoub S. Middle East respiratory syndrome coronavirus (MERS-CoV): what lessons can we learn? *Journal of Hospital Infection*. 2015;91:188–196. doi:10.1016/j.jhin.2015.08.002.

World Health Organization. MERS. Situation update. September 2019. https://www.who.int/health-topics/middle-east-respiratory-syndrome-coronavirus-mers#tab=tab_1. Accessed March 9, 2022.

COVID-19

Boggiano C, Eisinger RW, Lerner AM, et al. Update on and future directions for use of anti-SARS-CoV-2 Antibodies: National Institutes of Health summit on treatment and prevention of COVID-19. *Annals of Internal Medicine*. 2022;175(1):119–128. doi:10.7326/M21-3669.

Borczuk AC, Salvatore SP, Seshan SV, et al. COVID-19 pulmonary pathology: a multi-institutional autopsy cohort from Italy and New York City. *Modern Pathology*. 2020;33:2156–2168. doi:10.1038/s41379-020-00661-1.

Garduno-Soto M, Choreno-Parra JA, Cazarin-Barrientos J. Dermatological aspects of SARS-CoV-2 infection: mechanisms and manifestations. *Archives of Dermatological Research*. 2021;313(8):611–622. doi:10.1007/s00403-020-02156-0.

Garg M, Maralakunte M, Garg S, et al. The conundrum of "long-COVID-19": a narrative review. *International Journal of General Medicine*. 2021;14:2491–2506. doi:10.2147/IJGM.S316708.

Ghayda R, Hwa Lee K, Joo Han Y, et al. The global case fatality rate of coronavirus disease 2019 by continents and national income: a meta-analysis. *Journal of Medical Virology*. 2022:1–12. doi:10.1002/jmv.27610.

Junqueira C, Crespo A, Ranjbar S, et al. FcγR-mediated SARS-CoV-2 infection of monocytes activates inflammation. *Nature*. 2022. doi:10.1038/s41586-022-04702-4.

Lauer SA, Grantz KH, Bi Q, et al. The incubation period of coronavirus disease 2019 (COVID-19) from publicly reported confirmed cases: estimation and application. *Annals of Internal Medicine*. 2020;172(9):577–583. doi:10.7326/M20-0504.

Li H, Xue Q, Xu X. Involvement of the nervous system in SARS-CoV-2 infection. *Neurotoxicity Research*. 2020;38:1–7. doi:10.1007/s12640-020-00219-8.

Madjid M, Safavi-Naeini P, Solomon SD. Potential effects of coronaviruses on the cardiovascular system: a review. *JAMA Cardiology*. 2020;5(7):831–840. doi:10.1001/jamacardio.2020.1286.

Malieckal DA, Uppal NN, Ng JH, et al. Electrolyte abnormalities in patients hospitalized with COVID-19. *Clinical Kidney Journal*. 2021;14(6):1704–1707. doi:10.1093/ckj/sfab060.

Marshall M. COVID's toll on smell and taste: what scientists do and don't know. *Nature*. January 14, 2021. 589:342–343. doi:10.1038/d41586-021-00055-6.
Mesquita R da Rosa, Silva Junioer LCF, Santana FMS, et al. Clinical manifestations of COVID-19 in the general population: systematic review. *Wiener klinische Wochenschrift (The Central European Journal of Medicine)* 2021;133(7–8):377–382. doi:10.1007/s00508-020-01760-4.
Vessely MD, Perkins SH. Caution in the time of rashes and COVID-19. *Journal of the American Academy of Dermatology*. 2020;83(4):E321–E322. doi:10.1016/j.jaad.2020.07.026.
Zheng Y-Y, Ma Y-T, Zhang J-Y, et al. COVID-19 and the cardiovascular system. *Nature Reviews Cardiology* 2020;17:259–260. doi:10.1038//s41569-020-0360-5.

Coronavirus Comparisons

Madjid M, Safavi-Naeini P, Solomon SD, et al. Potential effects of coronaviruses on the cardiovascular system: a review. *JAMA Cardiology*. 2020;5(7):831–840. doi:10.1001/jamacardio.2020.1286.
Mann R, Perisetti A, Gajendran M, et al. Clinical characteristics, diagnosis, and treatment of major coronavirus outbreaks. *Frontiers in Medicine*. 2020;7:581521. doi:10.3389/fmed.2020.581521.
Ng Kee Kwong KC, Mehta PR, Shukla G, et al. COVID-19, SARS and MERS: a neurological perspective. *Journal Clinical Neuroscience*. 2020;77:13–16. doi:10.1016/j.jocn.2020.04.124.
Noroozi R, Branicki W, Pyrc K, et al. Altered cytokine levels and immune responses in patients with SARS-CoV-2 infection and related conditions. *Cytokine*. 2020;133:155143. doi:10.1016/j.cyto.2020.155143.
O'Sullivan O. Long-term sequelae following previous coronavirus epidemics. *Clinical Medicine*. 2021;21(1):e68–e70. doi:10.7861/clinmed.2020-0204.
Rogers JP, Chesney E, Oliver D, et al. Psychiatric and neuropsychiatric presentations associated with severe coronavirus infections: a systematic review and meta-analysis with comparison to the COVID-19 pandemic. *Lancet Psychiatry*. 2020;7(7):611–627. doi:10.1016/S2215-0366(20)30203–0.
Rongioletti F. SARS-CoV, MERS-CoV and COVID-19: what differences from a dermatological viewpoint? *Journal of the European Academy of Dermatology and Venereology*. 2020;34(10):e581–e582. doi:10.1111/jdv.16738.
Santacroce L, Charitos IA, Carretta DM, et al. The human coronaviruses (hCoVs) and the molecular mechanisms of SARS-CoV-2 infection. *Journal of Molecular Medicine*. 2021;99:93–106. doi:10.1007/s00109-020-02012-8.
Zhang X-Y, Huang H-J, Zhuang D-L, et al. Biological, clinical and epidemiological features of COVID-19, SARS and MERS and AutoDock simulation of ACE2. *Infectious Diseases of Poverty*. 2020;9:99. doi:10.1186/s40249-020-00691-6.
Zhao M, Wang M, Zhang J, et al. Advances in the relationship between coronavirus infection and cardiovascular diseases. *Biomedicine and Pharmacotherapy*. 2020;127:110230. doi:10.1016/j.biopha.2020.110230.

Appendix 3b

Cui J, Li F, Shi Z-L. Origin and evolution of pathogenic coronaviruses. *Nature Reviews Microbiology*. 2019;17(3):181–192. doi:10.1038/s41579-018-0118-9.
Forni D, Cagliani R, Clerici M, et al. Molecular evolution of human coronavirus genomes. *Trends in Microbiology*. 2017;25(1):35–48. doi:10.1016/j.tim.2016.09.001.

Guruprasad L. Human coronavirus spike protein-host receptor recognition. *Progress in Biophysics and Molecular Biology*. 2021;161:39–53. doi:10.1016/j.pbiomolbio.2020.10.006.

Hamre D, Beem M. Virologic studies of acute respiratory disease in young adults. V. Coronavirus 229E infections during six years of surveillance. *American Journal of Epidemiology*. 1972;96(2):94–106. doi:10.1093/oxfordjournals.aje.a121445.

Jaiswal NK, Saxena SK. (2020) Classical coronaviruses. In: Saxena S., ed. *Coronavirus disease 2019 (COVID-19): epidemiology, pathogenesis, diagnosis, and therapeutics*. Singapore: Springer; 2020. doi:10.1007/978-981-15-4814-7_12.

Laue M, Kauter A, Hoffmann T, et al. Morphometry of SARS-CoV and SARS-CoV-2 particles in ultrathin plastic sections of infected vero cell cultures. *Scientific Reports*. 2021;11:3515. doi:10.1038/s41598-021-82852-7.

Mann R, Perisetti A, Gajendran M, et al. Clinical characteristics, diagnosis, and treatment of major coronavirus outbreaks. *Frontiers in Medicine*. 2020;7:581521. doi:10.3389/fmed.2020.581521.

McIntosh K, Becker WB, Chanock RM. Growth in suckling-mouse brain of "IBV-like" viruses from patients with upper respiratory tract disease. *Proceedings of the National Academy of Sciences*. 1967;58(6):2268–2273. doi:10.1073/pnas.58.6.2268.

McIntosh K, Dees JH, Becker WB, et al. Recovery in tracheal organ cultures of novel viruses from patients with respiratory disease. *Proceedings of the National Academy of Sciences*. 1967;57(4):933–940. doi:10.1073/pnas.57.4.933.

Muller MA, Corman VM, Jores J, et al. MERS coronavirus neutralizing antibodies in camels, Eastern Africa, 1983–1997. *Emerging Infectious Diseases*. 2014;20(12):2093–2095. doi:10.3201/eid2012.141026.

National Library of Medicine, National Center for Biotechnology Information. NCBI Virus. Nucleotide sequences. SARS. MERS. SARS-CoV-2. https://www.ncbi.nlm.nih.gov/labs/virus/vssi/#/. Accessed April 11, 2022.

Santacroce L, Charitos IA, Carretta DM, et al. The human coronaviruses (HCoVs) and the molecular mechanisms of SARS-CoV-2 infection. *Journal of Molecular Medicine*. 2021;99:93–106. doi:10.1007/s00109-020-02012-8.

Tyrrell DA, Bynoe MS. Some further virus isolations from common colds. *British Medical Journal*. 1961;1(5223):393–397. doi:10.1136/bmj.1.5223.393.

Wu Y, Zhao S. Furin cleavage sites naturally occur in coronaviruses. *Stem Cell Research*. 2021;50:102115. doi:10.1016/j.scr.2020.102115.

Index